Concepts and Problems in Physics

Dr. Sanjay Kumar

LINEAR MOMENTUM, COLLISION & IMPULSE

FOR JEE (MAIN & ADVANCED) & NEET

Dr. Sanjay Kumar

M.Tech, PhD CMJU Meghalaya
Managing Director, Quanta Classes Lucknow, Mo. 9453763058
Ex. Sr. Faculty of Physics: Imagine Point Kanpur, Jain Classes Jhansi, Bansal Classes MP, TATA Aarambh Engineering and Medical Simplified Lucknow.

Copyright © 2020 Sanjay Kumar
All rights reserved.
No part of this book may be reproduced or distributed in any form or by any means, electronic, mechanical, photocopying, recording, or otherwise or stored in a database or retrieval system without the prior written permission of the author.

To JEE (MAIN & ADVANCED) & NEET aspirants
With the hope that this work will stimulate
an interest in Physics
and provide an acceptable guide to its understanding.

Online/Offline

Physics Classes

Dr. Sanjay Kumar

JEE(Main+Advanced)/NEET/Foundation (IX–XII)

Quanta Classes: K 423A Sector K Ashiyana Colony Lucknow, Mo. +919453763058

Email: spphysicsworld@gmail.com

PREFACE

This text book is the product of more than twenty years of teaching and innovation experience in physics for JEE and NEET aspirants. It is primarily intended for students who are preparing for the entrance tests of IIT-JEE/NEET/AIIMS and other esteemed colleges in same fields. This text is equally useful to the students preparing for their school exams.

Our main goals in writing this text book are
- to present the basic concepts and principles of physics that students need to know for their competitive exams.
- to provide a balance of quantitative reasoning and conceptual understanding, with special attention to concepts that have been causing difficulties to student in understanding the concepts.
- to develop students' problem-solving skills and confidence in a systematic manner.
- to motivate students by integrating real-world examples that build upon their everyday experiences.

Main Features of the Book-
1. Every concept is up to the mark and given in student friendly language with various solved problems. The solution is provided with problem solving approach and discussion.
2. Checkpoint questions have been added to applicable sections of the text to allow students to pause and test their understanding of the concept explored within the current section. The answers and solutions to the Checkpoints are given in answer keys, at the end of the chapter, so that students can confirm their knowledge without jumping too quickly to the provided answer.
3. Special attention is given to all tricky topics (like- how and when to apply conservation of linear momentum, impulse momentum theorem, series collisions, motion of centre of mass, inelastic, elastic and super elastic collisions, etc.) so that student can easily solve them with fun.
4. To test the understanding level of students, multiple choice questions, conceptual questions, practice problems with previous years JEE Main and Advanced problems are provided at the end of the whole discussion. Number of dots indicates level of problem difficulty. Straightforward problems (basic level) are indicated by single dot (•), intermediate problems (JEE mains and NEET level) are indicated by double dots (••), whereas challenging problems (advanced level) are indicated by thee dots (•••). Answer keys with hints and solutions are provided at the end of the chapter.

We have kept these goals in mind while developing the main themes of our physics book.

Dr. Sanjay Kumar

This page intentionally left blank

CONTENTS

1. **LINEAR MOMENTUM (LM)** 1
2. **NEWTON'S SECOND LAW IN TERMS OF LINER MOMENTUM** 1
   2.1. MOMENTUM WITHOUT MASS 1
3. **LINEAR IMPULSE** 1
   3.1. THE IMPULSE APPROXIMATION 2
   3.2. HOW TO APPLY LINEAR IMPULSE-MOMENTUM THEOREM IN PROBLEM SOLVING ... 2
   3.3. DIFFERENCE BETWEEN IMPULSE AND WORK 3
4. **SERIES OF COLLISIONS** 8
5. **CHECKPOINT 1** 10
6. **CONSERVATION OF LINEAR MOMENTUM & ITS APPLICATIONS** 10
   6.1. CONSERVATION OF LINEAR MOMENTUM 10
   6.2. APPLICATIONS 11
   6.3. CONSERVATION OF LINEAR MOMENTUM OF A BODY IN A COLLISION AT SUDDEN TURN OVER A SMOOTH SURFACE 13
7. **CHECKPOINT 2** 15
8. **SYSTEM OF VARIABLE MASS** 15
   8.1. SYSTEMS OF INCREASING MASS 15
9. **CHECKPOINT 3** 18
10. **SYSTEMS OF DECREASING MASS; ROCKET PROPULSION** 19
    10.1. MULTI STAGE ROCKET 20
11. **CHECKPOINT 4** 21
12. **CHAIN RELATED PROBLEMS** 22
    12.1. FORCE EXERTED BY FREE FALL OF CHAIN ON A SURFACE 22
13. **CHECKPOINT 5** 24
14. **CENTRE OF MASS (CM)** 26
15. **SYSTEMS OF PARTICLES** 26
    15.1. TWO PARTICLES 26
    15.2. MANY PARTICLES 27
16. **CM OF EXTENDED, CONTINIOUS OBJECTS** 27
17. **OPTIONAL DERIVATIONS** 30
    17.1. CM OF A HALF RING 30
    17.2. CM OF A HALF DISC 31
    17.3. CM OF A SOLID CONE 31

17.4. CM OF A HOLLOW CONE 32
17.5. CM OF HALF SHELL 32
18. THE CENTRE OF MASS AFTER REMOVAL OF A PART OF A BODY 32
19. MOTION OF THE CENTER OF MASS 33
20. FORCE AND MOMENTUM 33
21. SHIFT IN CM OF TWO PARTICLE SYSTEM IN ABSENCE OF EXTERNAL FORCE 34
22. CM SHIFT METHOD (DIRECT METHOD) 35
23. C-FRAME 35
24. KINETIC ENERGY OF A SYSTEM 35
25. A SYSTEM OF TWO PARTICLES 36
26. ANALYSIS OF SYSTEM OF TWO MASSES IN CM FRAME 36
27. CHECKPOINT 6 42
28. COLLISIONS 43
   28.1. CONSERVATION APPROACH IN A COLLISION 43
      28.1.1. CONSERVATION OF MOMENTUM DURING A COLLISION 43
      28.1.2. CONSERVATION OF KINETIC ENERGY DURING A COLLISION 44
   28.2. MATHEMATICAL ANALYSIS OF COLLISION 44
29. LAWS OF COLLISION 44
   29.1. HEAD ON (OR DIRECT) COLLISION (OR IMPACT) 44
      29.1.1. CONSERVATION OF MOMENTUM 44
      29.1.2. CONSERVATION OF KINETIC ENERGY 45
         29.1.2.1. ELASTIC COLLISION 45
      29.1.3. CALCULATION OF FINAL VELOCITIES AFTER ELASTIC COLLISION 45
         29.1.3.1. INELASTIC COLLISION 45
         29.1.3.2. SUPER-ELASTIC COLLISION 45
   29.2. NEWTON'S LAW FOR COLLISION 46
      29.2.1. CONCEPT OF COEFFICIENT OF RESTITUTION 46
      29.2.2. CALCULATION OF VELOCITIES AFTER DIRECT (OR HEAD ON) IMPACT BY USING NEWTON'S LAW OF RESTITUTION 46
30. CHECKPOINT 7 47
31. COLLISION OF A BALL WITH GROUND 48
   31.1. GENERALIZATION 48
      31.1.1. *Calculation of height after nth impact* 48
      31.1.2. *Calculation of velocity after nth impact* 48
      31.1.3. *Total distance covered by ball* 48
32. MISCONCEPTION 50
33. OBLIQUE OR INDIRECT COLLISION 50
34. INDIRECT IMPACT OF A BODY WITH A FIXED PLANE 51
35. IMPORTANT POINTS 53

36.	EXPLOSIONS AND CRASH-LANDINGS	54
37.	CHECKPOINT 8	55
38.	QUESTIONS AND EXERCISES	57
38.1.	CONCEPTUAL QUESTIONS (CQs)	57
38.2.	PRACTICE PROBLEMS (PPPs)	58
38.3.	MULTIPLE-CHOICE ASSIGNMENTS	61
38.3.1.	LEVEL 1	61
1.1.1.1.	CONSERVATION OF LINEAR MOMENTUM	61
1.1.1.2.	CENTRE OF MASS	63
1.1.1.3.	COLLISION	63
38.3.2.	LEVEL 2	66
38.3.3.	LEVEL 3	69
38.4.	PREVIOUS YEARS PROBLEMS	70
38.4.1.	●●SECTION-A (JEE MAIN)	70
38.4.2.	●●●SECTION - B [ADVANCED]	73
38.5.	MISCELLANEOUS QUESTIONS	75
39.	ANSWER KEYS AND SOLUTIONS	78
39.1.	CHECKPOINT 1	78
39.2.	CHECKPOINT 2	79
39.3.	CHECKPOINT 3	79
39.4.	CHECKPOINT 4	79
39.5.	CHECKPOINT 5	79
39.6.	CHECKPOINT 6	79
39.7.	CHECKPOINT 7	79
39.8.	CHECKPOINT 8	79
39.9.	CONCEPTUAL QUESTIONS	79
39.10.	PROBLEMS	81
39.11.	MULTIPLE-CHOICE ASSIGNMENTS	81
39.11.1.	LEVEL 1	81
39.11.2.	LEVEL 2	81
39.11.3.	LEVEL 3	81
39.11.4.	LEVEL 3	81
39.11.4.1.	SECTION A	81
39.11.4.2.	SECTION B	81
39.12.	MISCELLANEOUS QUESTIONS	82
39.13.	PRACTICE PROBLEMS	82

Online/Offline

Physics Classes

Dr. Sanjay Kumar

JEE(Main+Advanced)/NEET/Foundation (IX–XII)

Quanta Classes: K 423A Sector K Ashiyana Colony Lucknow, Mo. +919453763058

Email: spphysicsworld@gmail.com

LINEAR MOMENTUM, IMPULSE AND COLLISIONS

1. LINEAR MOMENTUM (LM)

Momentum of a body is defined to be the product of its mass m and velocity $\vec{v}$. It is a vector quantity and is represented by,
$$\vec{p} = m\vec{v}$$
Momentum depends on reference frame as velocity is frame dependent.
Unit: SI Unit of linear momentum is kg.m/s.
Kinetic energy and momentum are different
Kinetic energy and momentum are different quantities, even though both depend on mass and speed. Kinetic energy is a scalar, meaning it does not depend on direction. Two balls with the same mass and speed will always have the same kinetic energy. Momentum is a vector, so it *always* depends on direction. Two balls with the same mass and speed have opposite momentum if they are moving in opposite directions.
- ☞ KE and LM both are frame dependent.
- ☞ Internal forces can change the kinetic energy but not the linear momentum.

2. NEWTON'S SECOND LAW IN TERMS OF LINER MOMENTUM

By Newton's second law, the net external force $\vec{F}$ = rate of change of linear momentum
i.e., $\quad \vec{F} = \dfrac{d\vec{p}}{dt}$ ... (1)

i.e., $\quad \vec{F} = \dfrac{d(m\vec{v})}{dt} = m\dfrac{d\vec{v}}{dt} + \vec{v}\dfrac{dm}{dt}$

or $\quad \vec{F} = m\dfrac{d\vec{v}}{dt} + \vec{v}\dfrac{dm}{dt}$ ... (2)

Now, within the classical limit, m = constant, $\dfrac{dm}{dt} = 0$, therefore
$$\vec{F} = m\dfrac{d\vec{v}}{dt}$$
or $\quad \vec{F} = m\vec{a} \quad$ [as $\vec{a} = d\vec{v}/dt$] ... (3)

In physics, we also have situations (such as rocket motion) in which mass of the body changes but velocity of escape of mass is constant. In such situation as $\vec{v}$ = constant, therefore $\dfrac{d\vec{v}}{dt} = 0$, so equation (1) reduces to
$$\vec{F} = \vec{v}\dfrac{dm}{dt} \quad \text{... (4)}$$

2.1. MOMENTUM WITHOUT MASS

We have defined momentum for objects with mass. But momentum is a fundamental property of matter and energy. Light also carries momentum even though it is pure energy with no mass. The momentum of light depends on the energy of the light.

3. LINEAR IMPULSE

If a constant net force $\sum \vec{F}$ acts on a particle during a time interval Δt from t_1 to t_2, then by Newton's second law, we have
$$\sum \vec{F} = \dfrac{\overrightarrow{\Delta p}}{\Delta t} = \dfrac{\overrightarrow{p_2} - \overrightarrow{p_1}}{t_2 - t_1}$$
$$\Rightarrow \underbrace{\sum \vec{F}(t_2 - t_1)}_{\text{Linear Impulse}} = \overrightarrow{p_2} - \overrightarrow{p_1} \quad \text{... (1)}$$

The LHS of above equation (1), is called the **linear impulse** of the net force and it is denoted by $\vec{J}$ or $\vec{I}$.
$$\vec{J} = \sum \vec{F}(t_2 - t_1) = \sum \vec{F}\Delta t \quad \text{... (2)}$$
Thus, the product of the external force exerted on an object during a time interval and the time interval gives us a new quantity, the **impulse** of the force. Impulse is a vector quantity; its direction is the same as the net external force $\sum \vec{F}$.

From (3) and (1), we can write impulse in terms of linear momentum as
$$\vec{J} = \vec{p_2} - \vec{p_1} = m(\vec{v_2} - \vec{v_1}) \quad \text{... (3)}$$
(impulse–momentum theorem)

This expression is known as the **impulse-momentum theorem**. *It states that an impulse delivered to an object causes the object's momentum to change and the change in momentum during a time interval equals the impulse of the net force that acts on the object during that interval.* That is, the *effect* of an impulsive force is to change the object's momentum from $\vec{p_i}$ to
$$\vec{p_f} = \vec{p_i} + \vec{J}$$
From equation (3), it is clear that, the direction of impulse $\vec{J}$ is same as that of change in linear momentum.

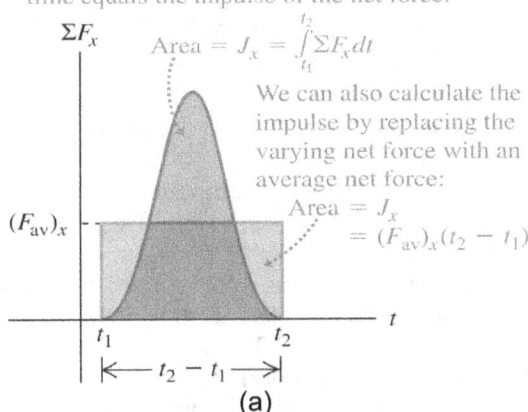
(a)

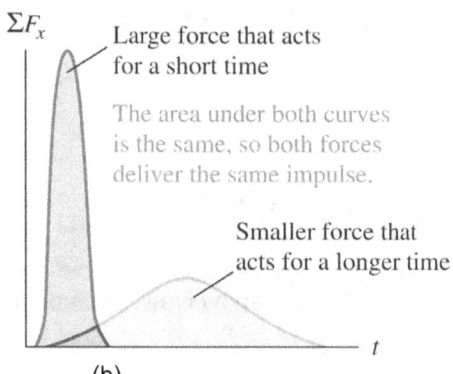

(b)
FIGURE.1 The meaning of the area under a graph of $\sum F_x$ versus t.

The impulse–momentum theorem also holds when forces are not constant. To see this, we integrate both sides of Newton's second law $\sum \vec{F} = \frac{d\vec{p}}{dt}$ over time between the limits t_1 and t_2.

$$\int_{t_1}^{t_2} \sum \vec{F}\, dt = \int_{t_1}^{t_2} \frac{d\vec{p}}{dt} dt = \int_{\vec{p}_1}^{\vec{p}_2} d\vec{p} = \vec{p}_2 - \vec{p}_1$$

The integral on the left is defined to be the impulse $\vec{J}$ of the net force $\sum \vec{F}$ during this interval:

$$\vec{J} = \int_{t_1}^{t_2} \sum \vec{F}\, dt = \vec{p}_2 - \vec{p}_1 \qquad \ldots (4)$$

(general definition of impulse)

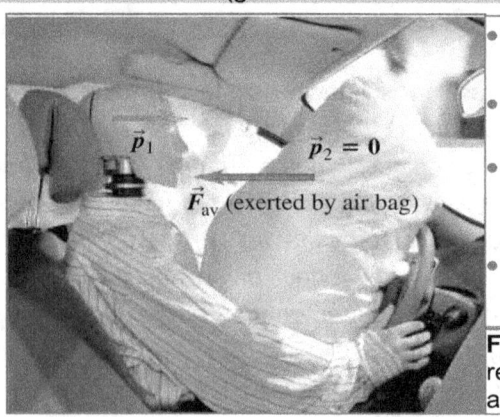

With this definition, the impulse–momentum theorem $\vec{J} = \vec{p}_2 - \vec{p}_1$, eq. (3), is valid even when the net force $\sum \vec{F}$ varies with time.

We can define an *average* net force $\vec{F}_{av}$ such that even when $\sum \vec{F}$ is not constant, the impulse is given by

$$\vec{J} = \vec{F}_{av}(t_2 - t_1) \qquad \ldots (5)$$

When $\sum \vec{F}$ is constant, $\sum \vec{F} = \vec{F}_{av}$ and Eq. (4) reduces to Eq. (2).

FIGURE 1a shows the x-component of net force as a function of time during a collision. This might represent the force on a soccer ball that is in contact with a player's foot from time t_1 to t_2. The x-component of impulse during this interval is represented by the red area under the curve between t_1 and t_2. This area is equal to the green rectangular area bounded by t_1, t_2 and $(F_{av})_x$, so $(F_{av})_x(t_2 - t_1)$ is equal to the impulse of the actual time-varying force during the same interval. Note that a large force acting for a short time can have the same impulse as a smaller force acting for a longer time if the areas under the force–time curves are the same (Fig. 1b). In this language, a car airbag (Fig. 2) provides the same impulse to the driver as would the steering wheel or the dashboard by applying a weaker and less injurious force for a longer time.

3.1. THE IMPULSE APPROXIMATION

When two objects interact during a collision or other brief interaction, (for example a collision between the bat and a ball), the forces *between* them are generally quite large. Other forces (gravitational, frictional etc.) may also act on the interacting objects, but usually these forces are *much* smaller than the interaction forces. We can reasonably ignore these small forces *during* the brief time of the impulsive force. Doing so is called the **impulse approximation.**

So, applying the impulse approximation means that Equation 3 reduces to just the impulses exerted by one colliding object on the other(s).

> ☞ In the impulse approximation, $\vec{p}_i$, $\vec{v}_i$ and $\vec{p}_f$, $\vec{v}_f$ are the momenta, velocities *immediately* before and *immediately* after the collision respectively.

- Impulse-momentum theorem:
  $\vec{J} = \vec{p}_2 - \vec{p}_1 = \vec{F}_{av}\Delta t$
- Impulse is the same no matter how the driver is brought to rest (so $\vec{p}_2 = 0$).
- Compared to striking the steering wheel, striking the air bag brings the driver to rest over a longer time interval Δt.
- Hence with an air bag, average force $\vec{F}_{av}$ on the driver is less.

FIGURE. 2 The impulse–momentum theorem explains how air bags reduce the chance of injury by minimizing the force on an occupant of an automobile.

3.2. HOW TO APPLY LINEAR IMPULSE-MOMENTUM THEOREM IN PROBLEM SOLVING

For problem solving, Eq. (4) will be rewritten in the form

$$\vec{p}_1 + \vec{J} = \vec{p}_2 \qquad \ldots (6)$$

i.e., $\quad m\vec{v}_1 + \int_{t_1}^{t_2} \sum \vec{F}\, dt = m\vec{v}_2$

In scalar form,

If $\vec{p}_1$ and $\vec{J}$ are in same direction, $\quad \left. \begin{array}{l} p_1 + J \\ p_1 - J \end{array} \right\} = p_2$
If $\vec{p}_1$ and $\vec{J}$ are in opposite directions,

or

$$\left. \begin{array}{l} \text{If } \vec{v}_1 \text{ is } \parallel \vec{F} \quad mv_1 + \int_{t_1}^{t_2} \sum F\, dt \\ \text{If } \vec{v}_1 \text{ is anti} \parallel \vec{F} \quad mv_1 - \int_{t_1}^{t_2} \sum F\, dt \end{array} \right\} = mv_2$$

which states that the initial momentum of the particle at

time t_1 plus the sum of all the impulses applied to the particle from t_1 to t_2 is equivalent to the final momentum of the particle at time t_2. These three terms are illustrated graphically on *the impulse and momentum diagrams* shown in Fig. 3. The two momentum diagrams are simply outlined shapes of the particle which indicate the direction and magnitude of the particle's initial and final momenta, $m\vec{v}_1$ and $m\vec{v}_2$. Similar to the free-body diagram, the impulse diagram is an outlined shape of the particle showing all the impulses that act on the particle when it is located at some intermediate point along its path.

If each of the vectors in Eq. 6 is resolved into its x, y, z components, we can write the following three scalar equations of linear impulse and momentum.

$$mv_{1x} + \int_{t_1}^{t_2} \Sigma F_x \, dt = mv_{2x}$$
$$mv_{1y} + \int_{t_1}^{t_2} \Sigma F_y \, dt = mv_{2y}$$
and $$mv_{1z} + \int_{t_1}^{t_2} \Sigma F_z \, dt = mv_{2z}$$

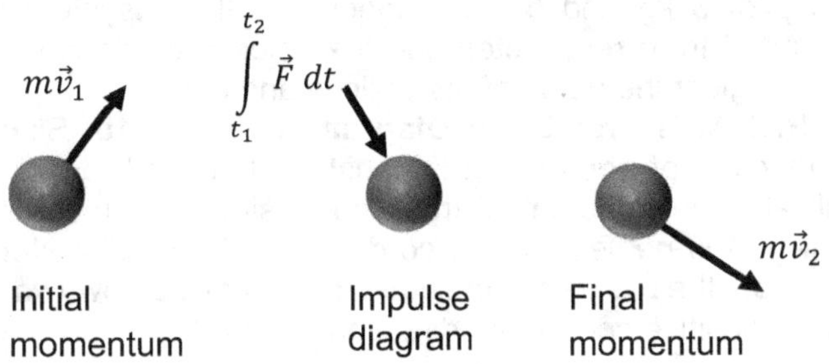

FIGURE 3

Initial momentum — Impulse diagram — Final momentum

You can find impulse by any of the following three methods:

1. by force method: $\vec{J} = \int_{t_1}^{t_2} \Sigma \vec{F} \, dt$

2. by momentum method: $\vec{J} = \int_{\vec{p}_1}^{\vec{p}_2} d\vec{p} = \vec{p}_2 - \vec{p}_1$

3. by graphical method: $J_x = $ area below the $F_x(t)$ graph. Similarly, we can find J_y and J_z.

Net impulse $\vec{J} = J_x\hat{\imath} + J_y\hat{\jmath} + J_z\hat{k}$

If we wish to know the average force, we can divide the impulse by the time interval during which the force is applied.

☞ During collision or impact, the impulsive forces produced between the colliding bodies, due to collision, are very strong, therefore in such situations, the impulse of weak forces, like gravity or friction, can be neglected. This approximation is called impulse approximation. Note that it is applicable in collision and explosions only.

IMPULSE-MOMENTUM THEOREM

$\vec{J} = \vec{p}_2 - \vec{p}_1 = \int_{t_1}^{t_2} \vec{F}_x dt = $ area below the $F_x(t)$ graph.

$\because \vec{F} = \frac{d\vec{p}}{dt}$ $\quad\therefore\quad d\vec{p} = \vec{F} dt$

Integrating both sides, we get

Impulse $\vec{J} = \int_{t_1}^{t_2} \vec{F} dt = \int_{\vec{p}_1}^{\vec{p}_2} d\vec{p} = \vec{p}_2 - \vec{p}_1$

(change in momentum)

From above equation, we can write: $\vec{p}_2 = \vec{p}_1 + \vec{J}$

In scalar form,

$$p_2 = \begin{cases} p_1 + J, & \text{if } \vec{p}_1 \text{ and } \vec{J} \text{ are in same direction} \\ p_1 - J, & \text{if } \vec{p}_1 \text{ and } \vec{J} \text{ are in opposite directions} \end{cases}$$

3.3. DIFFERENCE BETWEEN IMPULSE AND WORK

Impulse is a momentum transfer due to a force; work is an energy transfer due to a force.

	Impulse	Work
Definition	$\vec{F}\Delta t = p_2 - p_1$	$\vec{F}.\Delta\vec{r} = K_2 - K_1$
Vector or Scalar?	Vector	Scalar
Physical meaning	Momentum transfer	Energy transfer

Procedure for Analysis

The principle of linear impulse and momentum is used to solve problems involving force, time, and velocity, since these terms are involved in the formulation. For application it is suggested that the following procedure be used.

Free-Body Diagram
- Establish the x, y, z inertial frame of reference and draw the particle's free-body diagram in order to account for all the forces that produce impulses on the particle.
- The direction and sense of the particle's initial and final velocities should be established.
- If a vector is unknown, assume that the sense of its components is in the direction of the positive inertial coordinate(s). As an alternative procedure, draw the impulse and momentum diagrams for the particle as discussed in reference to Fig.3.

Impulse-Momentum Theorem
- In accordance with the established coordinate system, apply the principle of linear impulse and momentum, $m\vec{v}_1 + \int_{t_1}^{t_2} \Sigma \vec{F} dt = m\vec{v}_2$ If motion occurs in the x-y plane, the two scalar component equations can be formulated by either resolving the vector components of $\vec{F}$ from the free-body diagram, or by using the data on the impulse and momentum diagrams.
- Realize that every force acting on the particle's free-body diagram will create an impulse, even though some of these forces will do no work.
- Forces that are functions of time must be integrated to obtain the impulse. Graphically, the impulse is equal to the area under the force-time curve.

4 CONCEPTS AND PROBLEMS IN PHYSICS

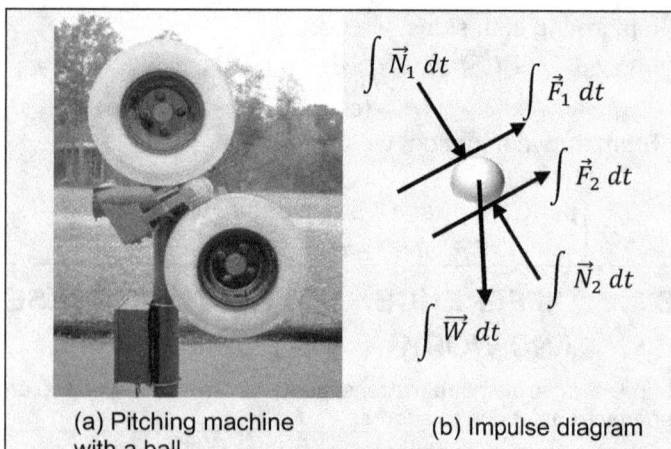

(a) Pitching machine with a ball

(b) Impulse diagram

FIGURE 4. As the wheels of the pitching machine rotate, they apply frictional impulses to the ball, thereby giving it a linear momentum. These impulses are shown on the impulse diagram. Here both the frictional and normal impulses vary with time. By comparison, the weight impulse is constant and is very small since the time Δt the ball is in contact with the wheels is very small.

EXAMPLE 1. The 100-kg stone shown in Fig. 1a is originally at rest on the smooth horizontal surface. If a towing force of 200 N, acting at an angle of 45°, is applied to the stone for 10 s, determine the final velocity and the normal force which the surface exerts on the stone during this time interval.

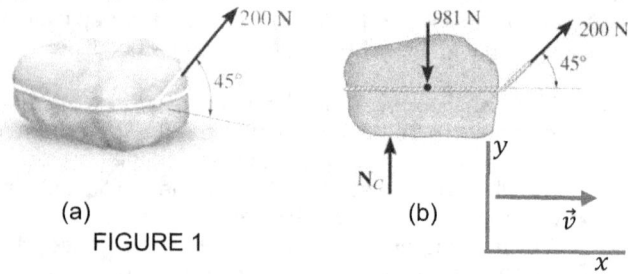

(a)   (b)
FIGURE 1

APPROACH This problem can be solved using the principle of impulse and momentum since it involves force, velocity, and time.

Free-Body Diagram. See Fig. 1b. Since all the forces acting are constant, the impulses are simply the product of the force magnitude and 10 s $[\vec{J} = \vec{F}(t_2 - t_1)]$. Note the alternative procedure of drawing the stone's impulse and momentum diagrams, Fig. 2.

Impulse-momentum Theorem Impulse momentum equations are-

Along X-axis: $mv_{1x} + \int_{t_1}^{t_2} \Sigma F_x \, dt = mv_{2x}$ ... (1)

Along Y-axis: $mv_{1y} + \int_{t_1}^{t_2} \Sigma F_y \, dt = mv_{2y}$ ... (2)

SOLUTION *Along X axis:* Substituting the values in equation (1), we get

$$0 + 200N \cos 45° (10) = (100 \, kg)v_2$$
$$v_2 = 14.1 \, m/s \quad \text{Ans.}$$

Along Y axis: Substituting the values in equation (1), we get

$$0 + N_C(10 \, s) - 981 \, N \, (10) + 200 \, N \sin 45° \, (10) = 0$$
$$N_C = 840 \, N \quad \text{Ans.}$$

NOTE: Since no motion occurs in the y direction, direct application of the equilibrium equation $\Sigma F_y = 0$, gives the same result for normal contact force N_C.

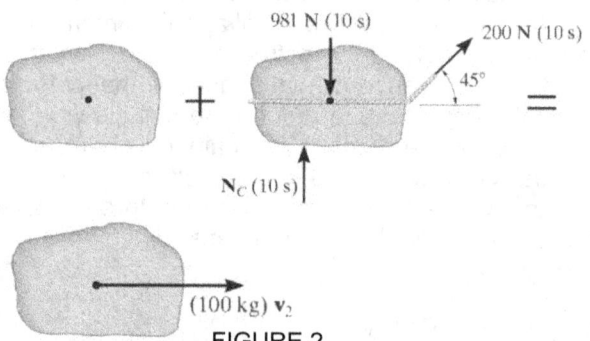

FIGURE 2

EXAMPLE 2. Blocks A and B shown in Fig. 1a have a mass of 3 kg and 5 kg, respectively. If the system is released from rest, determine the velocity of block B in 6 s. Neglect the mass of the pulleys and cord.

APPROACH Free-Body Diagram. See Fig. 1b. Since the weight of each block is constant, the cord tensions will also be constant. Furthermore, since the mass of pulley D is neglected, the cord tension $T_A = 2T_B$. Note that both the blocks are assumed to be moving downward in the positive coordinate directions, s_A and s_B.

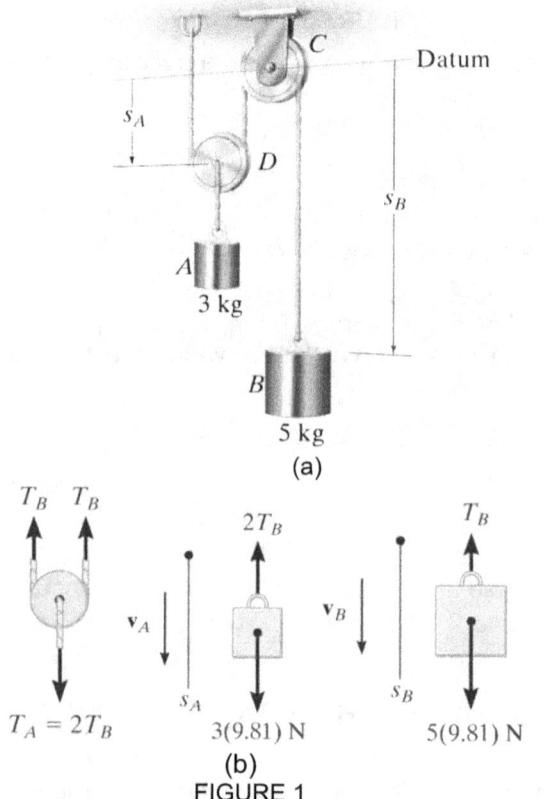

FIGURE 1

Kinematics. Since the blocks are subjected to dependent motion, the velocity of A can be related to that

of B by using the kinematic analysis. A horizontal datum is established through the fixed point at C, Fig. 1a, and the position coordinates, s_A and s_B, are related to the constant total length l of the vertical segments of the cord by the equation
$$2s_A + s_B = l$$
Taking the time derivative yields
$$2v_A = -v_B \qquad \ldots (1)$$
As indicated by the negative sign, when B moves downward A moves upward.

Impulse-momentum Theorem

Block A:
$$(+\downarrow) \quad mv_{A1} + \int_{t_1}^{t_2} \Sigma F_y \, dt = mv_{A2} \qquad \ldots (2)$$

Block B:
$$(+\downarrow) \quad mv_{B1} + \int_{t_1}^{t_2} \Sigma F_y \, dt = mv_{B2} \qquad \ldots (3)$$

SOLUTION Substituting, given values in (1), we get
$$0 - 2T_B(6\,s) + 3(9.81)\,N(6\,s) = (3\,kg)(v_{A2}) \qquad \ldots (4)$$
From equation (2), we get
$$0 + 5(9.81)N(6\,s) - T_B(6\,s) = (5\,kg)(v_{B2}) \qquad \ldots (5)$$
Using equation (1), in (4) and (5), we get
$$v_{B2} = 35.8 \text{ m/s} \downarrow \qquad \text{Ans.}$$
$$T_B = 19.2 \text{ N}$$

NOTE: Realize that the positive (downward) direction for v_A and v_B is consistent in Figs. 1a and 1b and in Equations 1 to 3. This is important since we are seeking a simultaneous solution of equations.

EXAMPLE 3. The loaded 150-kg skip is rolling down the incline at 4 m/s when a force F is applied to the cable as shown at time $t = 0$. The force F is increased uniformly with the time until it reaches 600 N at $t = 4$ s, after which time it remains constant at this value. Calculate (a) the time t' at which the skip reverses its direction and (b) the velocity v of the skip at $t = 8$ s. Treat the skip as a particle.

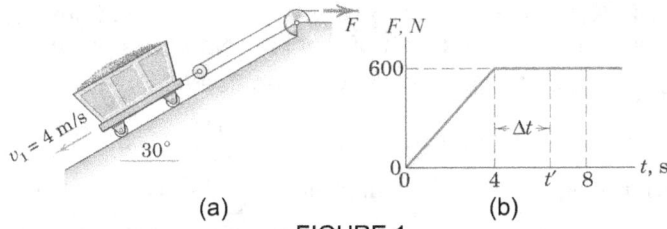

(a)   (b)

FIGURE 1

PART (a). APPROACH The stated variation of F with the time is plotted (FIGURE 1b), and the impulse momentum diagrams of the skip are drawn (FIGURE 2).
The skip reverses direction when its velocity becomes zero. We will assume that this condition occurs at $t' = 4 + \Delta t$ s.
To find t', we have to apply the impulse-momentum theorem consistently in the positive x-direction from $t = 0$ to $t = t'$.

FIGURE 2

Since the cable force on skip is $2F$, therefore to calculate its impulse from graph, we have to multiply area of F-t graph by 2.
The impulse of net cable force between $t = 0$ to $t = 4$ s
$= 2 \times$ area of triangle from $t = 0$ s to $t = 4$ s shown in FIGURE 1b
$= 2 \times \frac{1}{2} \times (600)(4)$ N.s (along +ve X direction)
The impulse of net cable force between $t = 4$ s to $t = t'$ s
$= 2 \times$ area of rectangle from $t = 4s$ to $t = t'$ shown in FIGURE 1b
$= 2 \times (600)\Delta t$ N.s (along +ve X direction)
The impulse of weight component from $t = 0$ s to $t = t'$ s
$= 150\,g \sin 30°\,(4 + \Delta t)$ N.s (along $-ve$ X direction)

SOLUTION By impulse momentum theorem, from $t = 0$ to $t = t'$, we have
$$mv_{1x} + \int \Sigma F_x \, dt = mv_{2x}$$
or
$$\underbrace{150(-4)}_{\text{Initial LM}} + \underbrace{2 \times \tfrac{1}{2}(4)(600)}_{\text{Impulse of cable force from } t=0 \text{ to } t=4\,s}$$
$$+ \underbrace{2 \times (600)\,\Delta t}_{\text{Impulse of cable force from } t=4 \text{ to } t=t'\,s}$$
$$- \underbrace{150\,g\sin 30°\,(4+\Delta t)}_{\text{Impulse of weight from } t=0 \text{ to } t=t'\,s} = \underbrace{150(0)}_{\text{Final LM}}$$
or $\quad 150(-4) + \tfrac{1}{2}(4)(2)(600) + 2(600)\,\Delta t -$
$150\,(9.81)\sin 30°\,(4+\Delta t) = 150(0)$
or $\quad \Delta t = 2.46\,s, \qquad t' = 4 + 2.46 = 6.46\,s$
$$\text{Ans.}$$

PART (b) APPROACH To find the velocity v of the skip at $t = 8$ s we apply the impulse momentum theorem to the entire 8-s interval gives
$$mv_{1x} + \int \Sigma F_x \, dt = mv_{2x}$$
$$150(-4) + 2 \times \tfrac{1}{2}(4)(600) + 2 \times (4)(600)$$
$$-150\,(9.81)\sin 30°\,(8) = 150(v_{2x})$$
$$v_{2x} = 4.76 \text{ m/s} \qquad \text{Ans.}$$
The same result is obtained by analysing the interval from t' to 8 s.

EXAMPLE 4. (i) A ball of mass 2 kg is moving towards a wall with speed 10 m/s. It strikes the wall perpendicularly and rebounds with the same speed. If time of contact between ball and wall is $1/100\,s$, then calculate the magnitude of average force on the ball.

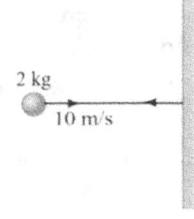

(ii) If the ball strikes to wall at an angle θ with normal and rebound with the same speed at the same angle. Calculate average force on the ball, if the time of impact is Δt.

(i) APPROACH In this problem- mass of the ball, velocity of ball before and after collision, and time of contact is given, therefore to calculate the magnitude of average force on the ball, we can use impulse-momentum theorem-

$$\vec{F}_{av}\Delta t = \vec{p}_2 - \vec{p}_1 = m\vec{v}_2 - m\vec{v}_1$$

SOLUTION $\vec{v}_2 = -10 \, m/s \, \hat{\imath}$, $\vec{v}_1 = +10 \, m/s \, \hat{\imath}$

By, impulse momentum theorem,

$$\vec{F}_{av}\Delta t = \vec{p}_2 - \vec{p}_1 = m\vec{v}_2 - m\vec{v}_1$$
$$= (2 \, kg)(-10 \, m/s \, \hat{\imath}) - (2 \, kg)(+10 \, m/s \, \hat{\imath})$$
$$= -40 \, kg.m/s \, \hat{\imath}$$

$$\vec{F}_{av} = \frac{-40 \, kg.m/s \, \hat{\imath}}{1/100 \, s} = -4000 \, kg.m/s^2 \, \hat{\imath}$$

here, $-ve$ sign shows that this force on the ball is acting along $-ve$ direction of x- axis (according to FIGURE)
or $\qquad |\vec{F}_{av}| = 4000 \, N$

(ii) APPROACH In this problem- mass of the ball, velocity of ball before and after collision, and time of contact is given, therefore to calculate the magnitude of average force on the ball, we can use impulse-momentum theorem-

$$\vec{F}_{av}\Delta t = \vec{p}_2 - \vec{p}_1 = m\vec{v}_2 - m\vec{v}_1$$

SOLUTION $\vec{v}_1 = v \cos\theta \, \hat{\imath} - v \sin\theta \, \hat{\jmath}$

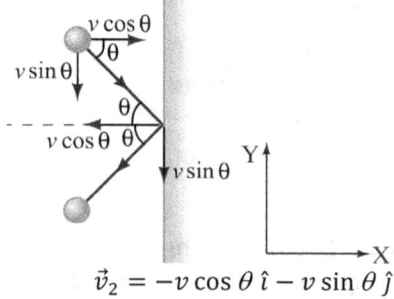

$$\vec{v}_2 = -v \cos\theta \, \hat{\imath} - v \sin\theta \, \hat{\jmath}$$

By, impulse momentum theorem,
$$\vec{F}_{av}\Delta t = \vec{p}_2 - \vec{p}_1$$
$$= m(\vec{v}_2 - \vec{v}_1)$$
$$= m[(-v\cos\theta \, \hat{\imath} - v \sin\theta \, \hat{\jmath}) - (v\cos\theta \, \hat{\imath} - v\sin\theta \, \hat{\jmath})]$$
$$= -2mv \cos\theta \, \hat{\imath}$$
$$\vec{F}_{av} = -\left(\frac{2mv\cos\theta}{\Delta t}\right)\hat{\imath}$$

here $-ve$ sign shows that this force on the ball is acting along $-ve$ direction of x- axis (according to FIGURE)

Magnitude of average force $|\vec{F}_{av}| = \frac{2mv\cos\theta}{\Delta t}$

EXAMPLE 5. A cart of mass $m_1 = 0.24 \, kg$ moves on a linear track without friction with an initial velocity of $0.17 \, m/s$. It collides with another cart of mass $m_2 = 0.68 \, kg$ that is initially at rest. The first cart carries a force probe that registers the magnitude of the force exerted by one cart on the other during the collision. The output of the force probe is shown in following figure. Find the velocity of each cart after the collision.

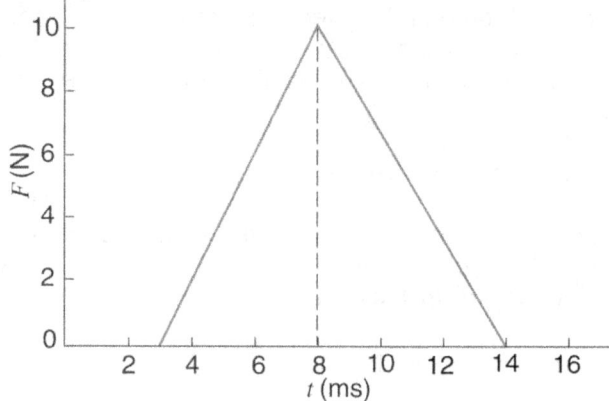

SOLUTION Impulse J = Area of the F-t graph on t axis
$= \frac{1}{2} \times 10 \times (14 - 3) \times 10^{-3} = 0.055 \, kg.m/s$.

This impulse is in opposite direction to the initial momentum of cart m_1.

Therefore, final momentum of cart m_1 is $p_{1f} = p_{1i} - J$
$\Rightarrow \qquad m_1v_1 = m_1u_1 - J = (0.24kg)(0.17 \, m/s) - 0.055$
$kg.m/s = -0.0142 \, kg.m/s$
$\Rightarrow \qquad v_1 = -\frac{0.0142}{0.24 \, kg} \, kg.m/s = -0.059 \, m/s = -5.9 \, cm/s.$

here negative sign shows that cart 1 reverses its direction of motion after collision.

As the impulse provides momentum to cart m_2 at rest, therefore, for cart m_2

$p_{2f} = p_{2i} + J = 0 + 0.055 \, kg.m/s$
$\Rightarrow m_2v_{2f} = 0.055 \, kg.m/s$
or $\qquad v_{2f} = \frac{0.055 \, kg.m/s}{0.68 kg} = 0.082 \, m/s = +8.2 \, cm/s$

☞ In problems involving impulsive forces: "Relative means after the event"

EXAMPLE 6. Force acting on a body is given by $F = 4t$ newton
(i) What is the impulse from $t = 1$ to $t = 3$ sec.
(ii) What is the average force in this interval

APPROACH Here, the force is given in terms of time 't', therefore for the first part of problem, we use impulse $J = \int_{t=1}^{3} F dt$ and for the second part, we use $F_{av} = \frac{J}{\Delta t}$

SOLUTION (i) impulse$= \int_{t=1}^{3} F dt = 4\left[\frac{t^2}{2}\right]_1^3 = 16 \, N.s$

(ii) average force in this interval $= \frac{J}{time} = \frac{16}{2} = 8 \, N$

☞ Slope of tangent in momentum-time graph will give force.

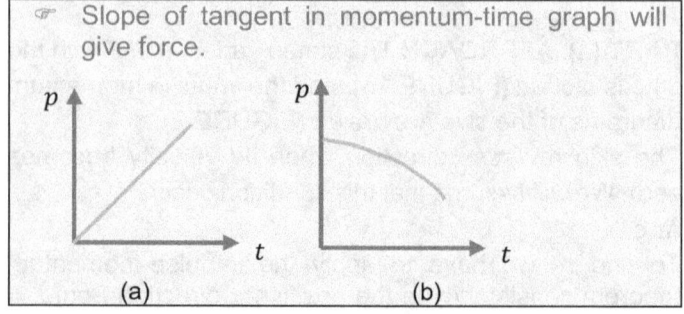

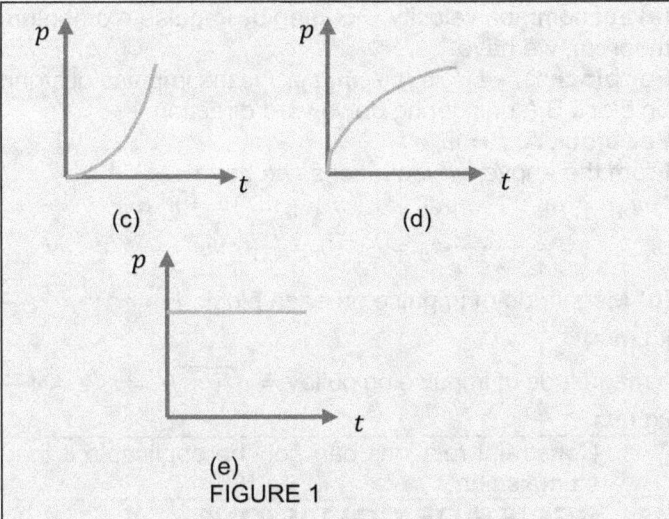

(a) F = constant (+ve) (b) F = increasing from zero to −ve value (c) F = increasing from zero to positive value (d) F = +ve decreasing (e) $F = 0$

For $0 < t < t_0$, $F = \frac{dp}{dt}$ = slope of tangent to the curve = +ve and it is decreasing up to $t = t_0$. At $t = t_0$, the slope of tangent to the curve is zero, therefore at $t = t_0$, the force $F = \frac{dp}{dt} = 0$. After t_0, the slope of tangent to the curve is negative and increasing with time, therefore the force is negative and increasing.

EXAMPLE 7. A pan being connected by a string passing over pulley counter balances a block of mass M. A mass m falls on the pan at rest from a height h from it and sticks to it. What is the speed of the block and the pan soon after the body hits the pan. Find the impulsive tension in the string.
(a) Connecting the pan and M.
(b) Connecting the pulley and the ceiling.

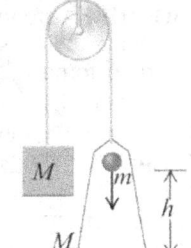

APPROACH During the impact, the impulses of the external forces are shown on the respective FBDs.

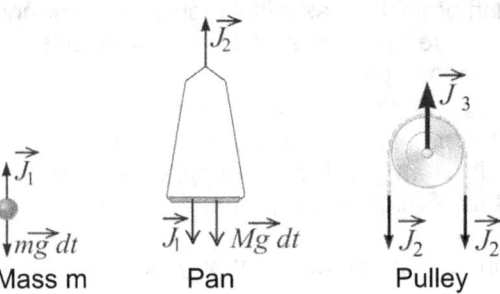

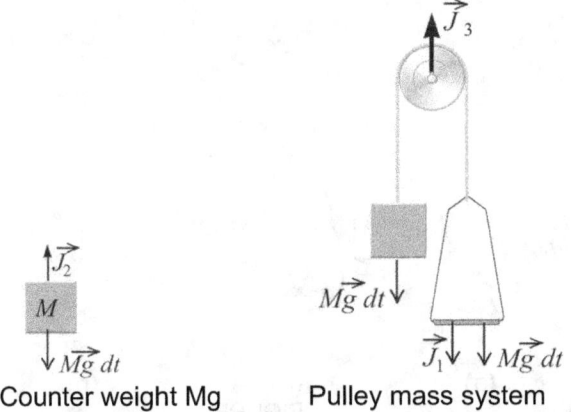

Counter weight Mg   Pulley mass system
FIGURE 1

For each FBD the impulse momentum theorem says **(Net Impulse = Change in momentum)**. During the impact the impulses of the weights mg dt, Mg dt can be neglected compare to the impulses of the forces due to impact.
$$mgdt, Mgdt \ll J_1, J_2, J_3$$
For any system (Here the pulley mass system), the momentum cannot be conserved by assuming that the net impulse $J_1 - J_3 \neq 0$
Now, we apply impulse-momentum theorem to different systems

SOLUTION Consider the FBD, of pan + mass m system
Momentum just before impact = $m\sqrt{2gh}$
Momentum after impact = $(m + M)v_1$
$$\therefore \quad m\sqrt{2gh} - J_2 = (m + M)v_1 \quad \ldots (1)$$
For the block M,
Momentum just before the impact = 0
Momentum just after the impact = Mv_2
$$\therefore \quad 0 + J_2 = Mv_2 \quad \ldots (2)$$
As the string is inextensible $v_1 = v_2$
From equation (1) and (2)
$-Mv_1 = (m + M)v_1 - m\sqrt{2gh}$
or $\quad (m + 2M)v_1 = m\sqrt{2gh}$
or $\quad v_1 = v_2 = \frac{m\sqrt{2gh}}{m+2M}$

Impulsive tension in the string connecting the pan and mass M, $J_2 = \frac{Mm\sqrt{2gh}}{m+2M}$

Impulsive tension in the string connecting the pulley and the ceiling, $J_3 = 2J_2 = \frac{2Mm\sqrt{2gh}}{m+2M}$.

EXAMPLE 8. A block of mass m and a pan of equal mass are connected by a string going over a smooth light pulley as shown in figure. Initially the system is at rest when a lump of putty of mass m falls on the pan and sticks to it. If the putty strikes the pan with a speed v find the speed with which the system moves just after the collision.

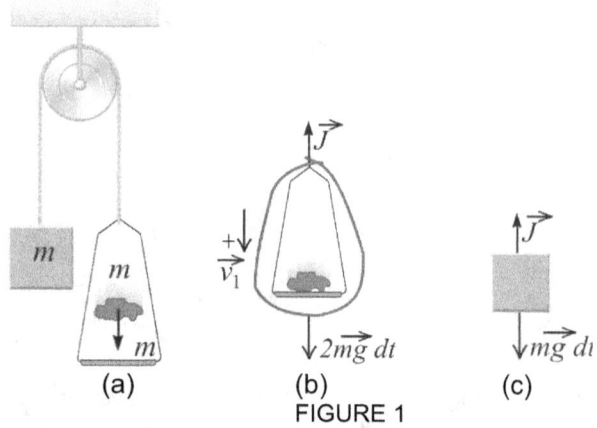
FIGURE 1

APPROACH Downward velocity of lump of putty of mass m just before the collision, is v. Suppose after collision the pan + lump of putty system (mass $= m + m = 2m$) moves with speed v_1, then for pan putty system:
As there is a sudden change in the speed of the block, the tension must change by a larger amount during the collision. Suppose $\vec{J}$ is the impulse of string on pan putty system, then (on neglecting the gravitational impulse)
$$mv - J = 2mv_1 \qquad \ldots (1)$$
(here we have considered downward direction as positive)
For block of mass m, we have
$$0 + J = mv_1 \qquad \ldots (2)$$
(we have neglected gravitational impulse)
SOLUTION From (1) and (2), we get
$0 = 3mv_1 - mv$ or $v_1 = v/3$

EXAMPLE 9. Two blocks A and B are joined by means of a slacked string passing over a massless pulley as shown in diagram. The system is released from rest and it becomes taut when B falls a distance 0.5 m, then
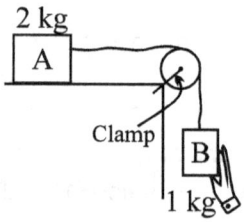
(a) Find the common velocity of two blocks just after string become taut.
(b) Find the magnitude of impulse on the pulley by the clamp during the small interval while string becomes taut.
SOLUTION Velocity of B just before the string is taut
$v_B = \sqrt{2gh} = \sqrt{10}$ m/s

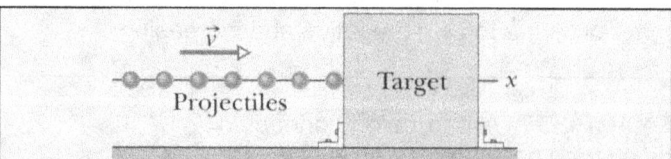

(a) Let common velocity $= v$, then by impulse momentum theorem, we have
For block B, $-J = m_B v - m_B v_B$, J is the impulse of string on block B (considering downward direction +ve)
For block A, $J = m_A v$
From the above two equations, we have
$-m_A v = m_B v - m_B v_B$ $\quad \therefore (m_A + m_B)v = m_B v_B$
or $\quad v = \dfrac{m_B}{m_A + m_B} v_B = \dfrac{1}{3}\sqrt{10} = \dfrac{\sqrt{10}}{3}$ m/s

(b) Magnitude of impulse on each block, $J = m_A v = 2\dfrac{\sqrt{10}}{3}$ kg.m/s
$\therefore$ magnitude of impulse on pulley $= \sqrt{J^2 + J^2} = J\sqrt{2} = 4\dfrac{\sqrt{5}}{3}$ kg.m/s

> Constraint relations can only be applicable if the string is tight.

4. SERIES OF COLLISIONS

Now let's consider the force on a body when it undergoes a series of identical, repeated collisions. For example, if we adjust a machine that fire tennis balls to fire them at a rapid rate directly at a wall. Each collision would produce a force on the wall, but that is not the force we are seeking. We want the average force F_{avg} on the wall during the bombardment—that is, the average force during a large number of collisions.

FIGURE 1 A steady stream of projectiles, with identical linear momenta, collides with a target, which is fixed in place. The average force F_{avg} on the target is to the right and has a magnitude that depends on the rate at which the projectiles collide with the target or, equivalently, the rate at which mass collides with the target.

In Fig. 1, a steady stream of projectile bodies, with identical mass m and linear momenta $m\vec{v}$, moves along x axis and collides with a target body that is fixed in place. Let n be the number of projectiles that collide in a time interval Δt. Because the motion is along only the x axis, we can use the components of the momenta along that axis. Thus, each projectile has initial momentum mv and undergoes a change Δp in linear momentum because of the collision. The total change in linear momentum for n projectiles during interval Δt is $n\Delta p$. The resulting impulse $\vec{J}$ on the target during Δt is along the x axis and has the same magnitude of $n\Delta p$ but is in the opposite direction. We can write this relation in component form as
$$J = -n\Delta p \qquad \ldots (1)$$
where the minus sign indicates that J and Δp have opposite directions.

Average Force. As linear impulse is given by,
$J = F_{avg}\Delta t$ ∴ Average force F_{avg} acting on the target during the collisions:
$$F_{avg} = \frac{J}{\Delta t} = -\frac{n}{\Delta t}\Delta p = -\frac{n}{\Delta t}m\Delta v \qquad \ldots (2)$$
This equation gives us F_{avg} in terms of $n/\Delta t$, the rate at which the projectiles collide with the target, and Δv, the change in the velocity of those projectiles.

Velocity Change. If the projectiles stop upon impact, then in Eq. 2 we can substitute, for Δv,
$$\Delta v = v_f - v_i = 0 - v = -v \qquad \ldots (3)$$
where $v_i(=v)$ and $v_f(=0)$ are the velocities before and after the collision, respectively. If, instead, the projectiles bounce (rebound) directly backward from the target with no change in speed, then $v_f = -v$ and we can substitute
$$\Delta v = v_f - v_i = -v - v = -2v \qquad \ldots (4)$$
In time interval Δt, an amount of mass $\Delta m = mn$ collides with the target. With this result, we can rewrite Eq. 2 as
$$\Delta F_{avg} = -\frac{\Delta m}{\Delta t}\Delta v. \qquad \ldots (5)$$
This equation gives the average force F_{avg} in terms of $\Delta m/\Delta t$, the rate at which mass collides with the target. Here again we can substitute for Δv from Eq. 3 or 4 depending on what the projectiles do.

EXAMPLE 10. A plastic disk of 50 g remains float in air when some particles strike to lower surface of the disc. Mass of each particle is 2g, and they strike to the disc with speed 2 m/s and

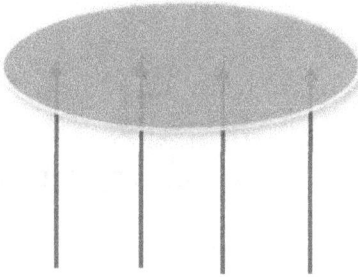

rebounds with the same speed. Calculate the number of particles strike to the disc per min.
APPROACH As the plastic disk floats in air when some particles strike to lower surface of the disc, therefore the average force exerted by particles is equal to the weight of the disc (Mg). i.e.,
$F_{avg} = -\frac{n}{\Delta t}\Delta p = Mg$.
SOLUTION Force exerted by particles $F_{avg} = Mg$
or $\qquad -\frac{n}{\Delta t}\Delta p = Mg$
or $\qquad -\frac{n}{\Delta t}m\Delta v = Mg$
or $\qquad \frac{(mn)(2v)}{60} = Mg$, here m is the mass of a striking particle.
or $\qquad n = \frac{Mg}{m(2v)} \times 60$
$n = \frac{50\times 10^{-3}\times 10}{(2\times 10^{-3})\times 2(2)} \times 60 = \frac{125}{2} \times 60 = 3750$

EXAMPLE 11. Jet of water strikes to a wall with speed v and rebounds with the same speed. Cross-sectional area of jet is A, and density of water is ρ. Calculate the force on the wall.

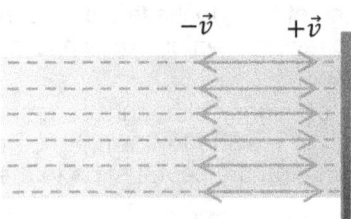

APPROACH This situation is also similar to series of collision, here the collision is continuous by jet of water. In this case the average force is given by
$$\Delta F_{avg} = -\frac{\Delta m}{\Delta t}\Delta v$$
SOLUTION The average force exerted by jet of water on wall
$F_{avg} = -\frac{\Delta m}{\Delta t}\Delta v = -\frac{\Delta m}{\Delta t}(-v-v) = 2v\frac{\Delta m}{\Delta t}$
As mass Δm = volume × density = $A\Delta l \rho$
∴ $\qquad F_{avg} = 2v\frac{\Delta m}{\Delta t} = 2v\left(\frac{\rho A \Delta l}{\Delta t}\right) = 2v\left(\rho A \frac{\Delta l}{\Delta t}\right) = 2v(\rho A v) = 2\rho A v^2$

EXAMPLE 12. WASHING A CAR: MOMENTUM CHANGE AND FORCE
(a) Water leaves a hose at a rate of with a speed of and is aimed at the side of a car, which stops it, Fig. 2. (That is, we ignore any splashing back.) What is the force exerted by the water on the car?
(b) If the water splashes back from the car in part (a), would the force on the car be larger or smaller?

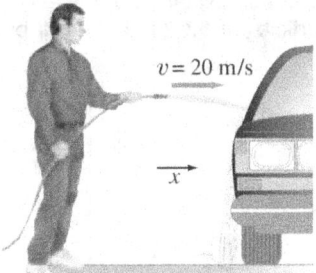

APPROACH The water leaving the hose has mass and velocity, so it has a momentum in the horizontal (x) direction, and we assume gravity doesn't pull the water down significantly. When the water hits the car, the water loses this momentum ($p_{final} = 0$). We use Newton's second law in the momentum form to find the force that the car exerts on the water to stop it. By Newton's third law, the force exerted by the water on the car is equal and opposite. We have a continuing process: 1.5 kg of water leaves the hose in each 1.0-s time interval. So, let us write $F = \Delta p/\Delta t$, where $\Delta t = 1.0\,s$ and $p_{initial} = mv_{initial} = (1.5\,kg)(20\,m/s)$.
SOLUTION The force (assumed constant) that the car must exert to change the momentum of the water is
$$F = \frac{\Delta p}{\Delta t} = \frac{p_{final} - p_{initial}}{\Delta t} = \frac{0 - 30\,kg.m/s}{1.0\,s} = -30\,N$$
The minus sign indicates that the force exerted by the car on the water is opposite to the water's original velocity.

The car exerts a force of 30 N to the left to stop the water, so by Newton's third law, the water exerts a force of 30 N to the right on the car.

NOTE Keep track of signs, although common sense helps too. The water is moving to the right, so common sense tells us the force on the car must be to the right.

(b) Larger (why?)

EXAMPLE 13. What is force time relation for $p = t^2 + 2t$ kg-m/s

SOLUTION $F = \frac{dp}{dt} = 2t + 2$ newton

5. CHECKPOINT 1

1. ••A glider of mass m is free to slide along a horizontal air track. It is pushed against a launcher at one end of the track. Model the launcher as a light spring of force constant k compressed by a distance x. The glider is released from rest. (a) Show that the glider attains a speed of $v = x\left(\frac{k}{m}\right)^{1/2}$. (b) Show that the magnitude of the impulse imparted to the glider is given by the expression $I = x(km)^{1/2}$. (c) Is more work done on a cart with a large or a small mass?

2. ••The magnitude of the net force exerted in the x direction on a 2.50 kg particle varies in time as shown in FIGURE P2. Find (a) the impulse of the force over the 5.00 s time interval, (b) the final velocity the particle attains if it is originally at rest, (c) its final velocity if its original velocity is $-2.00\ \hat{\imath}$ m/s, and (d) the average force exerted on the particle for the time interval between 0 and 5.00 s.

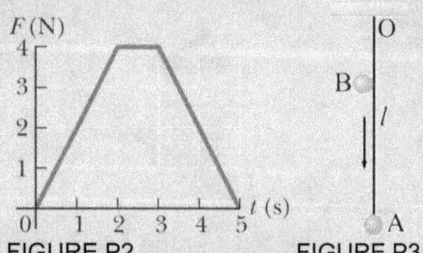

FIGURE P2   FIGURE P3

3. ••A small steel ball A is suspended by an inextensible thread of length $l = 1.5$ m from O (FIGURE P3). Another identical ball is thrown vertically downwards such that its surface remains just in contact with thread during downward motion and collides elastically with the suspended ball. If the suspended ball just completes vertical circle after collision, calculate the velocity of the falling ball just before collision. ($g = 10\ ms^{-2}$).

6. CONSERVATION OF LINEAR MOMENTUM & ITS APPLICATIONS

6.1. CONSERVATION OF LINEAR MOMENTUM

Since, $\vec{F} = \overrightarrow{dp/dt}$

If, $\vec{F} = 0$, then

$\frac{\overrightarrow{dp}}{dt} = 0$ or $\vec{p}$ = constant.

Thus, we can say that- if net external force acting on the particle or system of particles is zero then the total momentum of the particle or system of particles remain conserved.

☞ **Conservation of momentum means conservation of its components.** When you apply the conservation of momentum to a system, remember that momentum is a *vector* quantity. Hence you must use vector addition to compute the total momentum of a system (see adjoining figure). Using components is usually the simplest method.

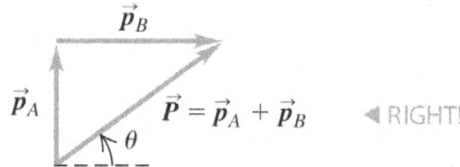

A system of two particles with
$p_A = 18$ kg · m/s   momenta in
$p_B = 24$ kg · m/s   different directions

You CANNOT find the magnitude of the total momentum by adding the magnitudes of the individual momenta!

$P = p_A + p_B = 42$ kg · m/s   ◀ WRONG

Instead, use vector addition:

$\vec{P} = \vec{p}_A + \vec{p}_B$ ◀ RIGHT!

☞ Before applying conservation of momentum to a problem, you must decide whether momentum *is* conserved! This will be true *only* if the vector sum of the external forces acting on the system of particles is zero. If this is not the case, you can't use conservation of momentum.

☞ In some ways the principle of conservation of momentum is more general than the principle of conservation of mechanical energy. For example, mechanical energy is conserved only when the internal forces are *conservative*—that is, when the forces allow two-way conversion between kinetic and potential energy—but conservation of momentum is valid even when the internal forces are *not* conservative.

☞ If an impulsive force acts on any object then its momentum along the direction of force cannot be conserved.

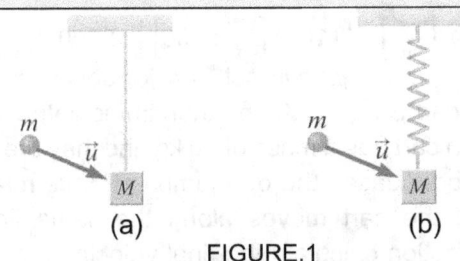
FIGURE.1

In Fig.1(a) momentum cannot be conserved in vertical direction just after collision because just after the collision, the tension (T) reaches to a very large value in a very small time and $\int_0^{\Delta t} T\, dt \neq 0$, while in (b) it can be conserved momentarily just at the time of collision because spring cannot change force immediately just after collision.

MOMENTUM CONSERVATION APPLIES TO A SYSTEM!

The momentum of an isolated system is conserved, but not necessarily the momentum of one particle within that system, because other particles in the system may be interacting with it. Apply conservation of momentum to an isolated system *only*.

6.2. APPLICATIONS

EXAMPLE 14. A girl of mass M standing on a frictionless floor, throws a ball of mass m with a speed u as shown in figure. If the velocity of girl, after he throws the ball is $\vec{v}$, then find the value of $\vec{v}$.

APPROACH Initially, the girl + ball system is at rest, so its initial linear momentum is zero. Since, there is no external force on girl + ball system, along horizontal direction, therefore linear momentum, of the system, in horizontal direction will remain conserved. Hence by applying conservation of LM, we can find the required velocity of girl.

SOLUTION Initial linear momentum of the girl + ball system, $\vec{p}_{\text{initial}} = 0$
Final linear momentum of the system,
$\vec{p}_{\text{final}} = -mu\hat{\imath} + M\vec{v}$, here $\hat{\imath}$ is the unit vector along $+ve$ direction of x axis.
As $\vec{p}_{\text{initial}} = \vec{p}_{\text{final}}$
$\vec{v} = \dfrac{m\hat{\imath}}{M}$

EXAMPLE 15. A ball of mass m moving with constant speed of u m/s is caught by a girl of mass M standing on a frictionless floor. If the velocity acquired by the girl is $\vec{v}$, then find $\vec{v}$.

APPROACH Since, there is no external force on girl + ball system along horizontal direction, therefore linear momentum of the system, in horizontal direction will be conserved. Hence by applying conservation of linear momentum, we can find the required velocity of girl.

SOLUTION $\vec{p}_{\text{initial}} = mu\hat{\imath}$, here $\hat{\imath}$ is the unit vector along $+ve$ direction of x axis.
$\vec{p}_{\text{final}} = (m+M)\vec{v}$
$\vec{p}_{\text{initial}} = \vec{p}_{\text{final}}$
$\vec{v} = \dfrac{mu\hat{\imath}}{(m+M)}$

EXAMPLE 16. A ball of mass 0.2 kg rests on vertical post of height 5 m. A bullet of mass 0.01 kg, travelling with velocity v m/s in horizontal direction, hits the centre of the ball. After the collision, the ball and the bullet travel independently. The ball hits the ground at a distance of 20 m and the bullet at a distance of 100 m from the foot of the post. The initial velocity v of the bullet is [Take $g = 10$ m/s^2.]   **[IIT 2011]**
(A) 250 m/s (B) $250\sqrt{2}$ m/s
(C) 400 m/s (D) 500 m/s

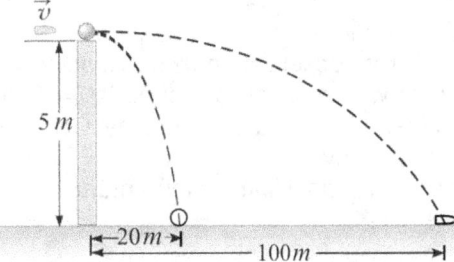

APPROACH After collision bullet and ball both of them

will initially move horizontally, therefore after collision their initial velocities in vertical direction will be zero. The height of vertical post is also given so, we can easily calculate their time of flight by applying $h = \frac{1}{2}gt^2$. As ranges of bullet and ball are given, therefore by applying equation of motion $s = ut + \frac{1}{2}at^2$, with acceleration $a = 0$, in horizontal direction, we can find their respective velocities after collision. Now, by applying conservation of linear momentum before and after collision we can easily find velocity of bullet before collision.

SOLUTION (D) The time of flight for the bullet is same as that of the ball and is given by
$h = \frac{1}{2}gt^2 \Rightarrow t = \sqrt{2h/g} = \sqrt{2 \times 5/10} = 1s$

Just after the collision, the velocity of bullet (v_1) is related to its range (R_1) by
$v_1 t = R_1 \qquad \Rightarrow v_1 = R_1/t = 100/1 = 100 \; m/s$
or $\qquad v_1 = 100 \; m/s$.

Similarly, the velocity of the ball is
$v_2 = R_2/t = 20/1 = 20 \; m/s$.

Consider the bullet and the ball together as a system. Along the direction of collision, there is no external force on the system. Hence, linear momentum of the system in the direction of collision is conserved. The linear momentum of the system before and after the collision are-
$p_i = m_1 v$
$p_f = m_1 v_1 + m_2 v_2$

By conservation of linear momentum, we have $p_i = p_f$
$\therefore \quad v = \frac{m_1 v_1 + m_2 v_2}{m_1} = \frac{(0.01)(100) + (0.2)(20)}{0.01} = 500 \; m/s$

EXAMPLE 17. In the following figure, a soldier with a rocket launcher is standing on a movable cart. Given that- mass of soldier + rocket launcher system $M = 90$ kg., mass of rocket $m = 10$ kg, velocity of system is $u = 5$ m/s and muzzle velocity of rocket $v_0 = 30$ m/s.

FIGURE 1. soldier + Rocket Launcher

(a) What will be soldier's & rocket's velocity after firing.
(b) Find energy of explosion

☞ Muzzle velocity = velocity of bullet w.r.t. Gun. Internal forces can change the KE but not Psystem.

SOLUTION Soldier fires rocket with muzzle velocity = v_0 Let velocity of rocket with respect to ground is v_2 & that of soldier + launcher system is v_1, then by COLM in horizontal direction, we have
Initial linear momentum = final linear momentum
$(m + M)u = Mv_1 + mv_2$
here, $v_2 = v_0 + u$
$\therefore \qquad (m + M)u = Mv_1 + m(v_0 + u)$
or $\qquad Mu - mV_0 = MV_1$

or $\qquad v_1 = \frac{Mu - mv_0}{M}$
or $\qquad v_1 = \frac{90 \times 5 - 10 \times 30}{90} = \frac{450 - 300}{90} = \frac{150}{90} = \frac{5}{3}$ m/s

Energy of explosion $= KE_f - KE_i = \left(\frac{1}{2}MV_1^2 + \frac{1}{2}mV_2^2\right) - \frac{1}{2}(m + M)u^2$

$= \frac{1}{2}(m + M)\left[\frac{Mu^2 + mv_0^2}{M} - u^2\right]$

$= \frac{1}{2}(m + M)\left[\frac{mv_0^2}{m}\right] = \frac{1}{2}(100)\left[\frac{10}{90} \times 900\right] = 5000$ J

EXAMPLE 18. In a circus act, a 4 kg dog is trained to jump from cart B cart to A and then immediately back to cart B. Each cart has a mass of 20 kg and they are initially at rest. In both cases the dog jumps at 6m/s relative to the cart. If the cart moves along the same line with negligible friction calculate the final velocity of each cart with respect to the floor.

APPROACH As there is no external force on dog and cart system, therefore, we apply conservation of linear momentum in, ground frame, in horizontal direction.

SOLUTION Given that, the magnitude of velocity of dog with respect to cart is $v_{DC} = 6 \; m/s$

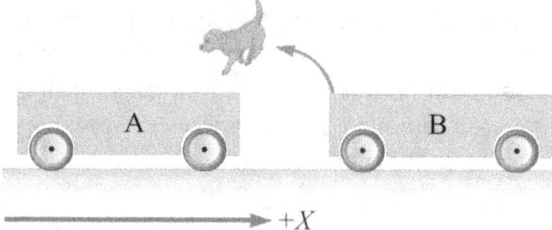

Jump of Dog from Cart B to A: When dog jumps from cart B to A,
$\vec{v}_{DB} = \vec{v}_D - \vec{v}_B$
$\vec{v}_D = \vec{v}_{DC} + \vec{v}_B$
$\vec{v}_D = -6\hat{\imath} + \vec{v}_B$
$\therefore$ By COLM, we have
$0 = m_D \vec{v}_D + m_B \vec{v}_B$
$\therefore \qquad 0 = m_D(-6\hat{\imath} + \vec{v}_B) + m_B \vec{v}_B$
or $\qquad 0 = 4(-6\hat{\imath} + \vec{v}_B) + 20\vec{v}_B$
or $\qquad \vec{v}_B = 1\hat{\imath}$

Thus, after jump, cart B start moving with velocity 1 m/s along positive direction of X axis
here, negative sign shows that this velocity is towards A. So, after first jump, the velocity of dog is $-5\hat{\imath}$.

Landing of Dog on Cart A:
When dog lands on cart A, both the dog and cart A will move with a common velocity $\vec{v}_{common}$ (say).

By COLM, we have $-5\hat{i} \times 4 = (20+4)\vec{v}_{common}$
$\vec{v}_{common} = -5/6\,\hat{i}$
In this case both will move with velocity $-5/6\,\hat{i}$ in negative direction of x axis.

Jump of Dog from Cart A to B:
Let after jump of dog from cart A, towards B, the velocity of cart A becomes $\vec{v}_A$, then
$$\vec{v}_D = 6\hat{i} + \vec{v}_A$$

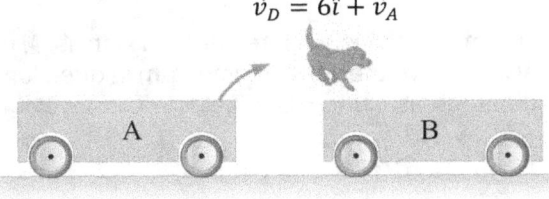

$-(24) \times \frac{5}{6}\hat{i} = (6\hat{i} + \vec{v}_A) \times 4 + \vec{v}_A \cdot 20$

$\vec{v}_A = -\frac{11}{6}\hat{i}$

Landing of Dog on Cart B:
Velocity of dog, just before landing on cart B, is $\vec{v}_D = -\frac{11}{6}\hat{i} + 6\hat{i} = \frac{25}{6}\hat{i}$ and velocity of cart B is the $\vec{v}_B = \hat{i}$

Suppose after landing on cart B, both start moving with common velocity $\vec{v}'_{common}$, then

$M_D \vec{v}_D + M_B \vec{v}_B = (M_B + M_D)\vec{v}'_{common}$

$\Rightarrow \quad 4\left(+\frac{25}{6}\hat{i}\right) + 20(\hat{i}) = 24\vec{v}'_{common}$

or $\quad 24\vec{v}'_{common} = \frac{220}{6}\hat{i}$

or $\quad \vec{v}'_{common} = \frac{55}{36}\hat{i}$ m/sec

So, the final velocities are: $v_A = \frac{11}{6}$ m/s (towards left),

$v_D = v_B = \frac{55}{36}$ m/s (towards right)

6.3. CONSERVATION OF LINEAR MOMENTUM OF A BODY IN A COLLISION AT SUDDEN TURN OVER A SMOOTH SURFACE

Suppose a small block of mass M slides on a frictionless surface of an inclined plane (FIGURE 1a). The angle of the incline suddenly changes from α to β at point B. If the block collides at B, then the direction of impulsive force due to collision on the block will be along the normal vector on the block at the surface BC, i.e., perpendicularly outward to the surface BC. Since the surface BC is smooth, therefore there will not be any frictional force along the surface BC and the effect of non-impulsive gravitational force can be neglected as compared to the impulsive collision force (impulse approximation). Thus, there is no impulsive external force on the block in a direction parallel to BC.

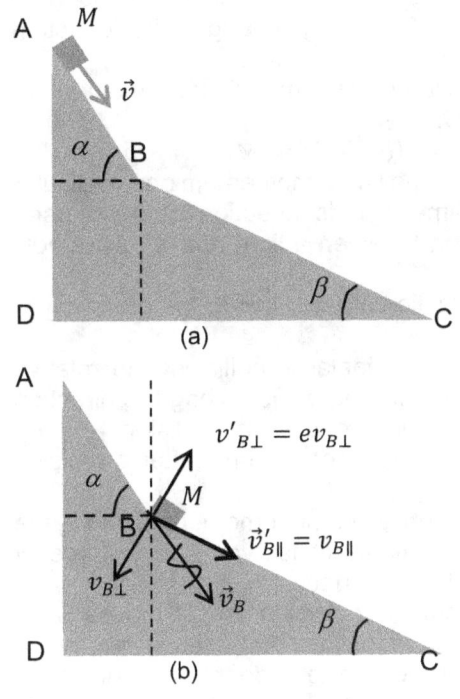

FIGURE 1

➤ **Calculation of velocity and linear momentum just after collision at B parallel and perpendicular to the surface BC**

1. Parallel to Surface BC
Since there is no force on the block parallel to the surface BC. Therefore, the component of linear momentum (and hence velocity) parallel to surface BC remains unchanged, i.e.,
$\vec{p}'_{BC\parallel} = \vec{p}_{BC\parallel}$.
or $\quad Mv'_{B\parallel} = Mv_{B\parallel}$
or $\quad v'_{B\parallel} = v_{B\parallel}$

2. Perpendicular to Surface BC
Since, there is a normal impulsive force on the block perpendicular to surface BC, therefore its velocity component perpendicular to surface BC get changed. If the velocity component of the block perpendicular to surface BC before the collision is $v_{B\perp}$ and the coefficient of restitution is e, then after collision the velocity component perpendicular to the plane BC is given by-
$v'_{B\perp} = ev_{B\perp}$ (in the direction of normal impulsive force i.e., perpendicularly outward to BC)
Here it is to be noted that the direction of $v'_{B\perp}$ is exactly opposite to $v_{B\perp}$.
In vector form we can write it as, $\vec{v}'_{B\perp} = -e\vec{v}_{B\perp}$
here negative sign shows that, after collision, the velocity component perpendicular to the plane BC reverse its direction. [Before collision it was perpendicularly inward to the surface BC but after collision it becomes

perpendicularly outward (FIGURE 1)].
For elastic collision, $e = 1$, therefore $\vec{v}'_{B\perp} = -\vec{v}_{B\perp}$ or $v'_{B\perp} = v_{B\perp}$
i.e., for elastic collision at B, the velocity component perpendicular to surface only changes its direction whereas its magnitude remains same.

Thus, the magnitude of the linear momentum perpendicular to BC,
$$\vec{p}'_{BC\perp} = M\vec{v}'_{B\perp} = M(-\vec{v}_{B\perp}) = -M\vec{v}_{B\perp} = -\vec{p}_{BC\perp}$$
i.e., the magnitude of the linear momentum perpendicular to BC remains same but its direction gets reversed (similar to a collision between a light and an extremely heavy ball).

For completely inelastic collision, $e = 0$, therefore $v'_{B\perp} = 0$.

That is, in a completely inelastic collision, the relative velocity between the colliding bodies along the direction of collision force become zero after the collision.

Therefore, the linear momentum after collision perpendicular to BC, $\vec{p}'_{BC\perp} = 0$

☞ If surface is a rough, then linear momentum along the surface cannot remains conserved due to force of friction parallel to the surface.

EXAMPLE 19. A small block of mass M moves on a frictionless surface of an inclined plane, as shown in the figure. The angle of the incline suddenly changes from $60°$ to $30°$ at point B. The block is initially at rest at A. Assume that collisions between the block and the incline are totally inelastic. [Take $g = 10$ m.s^2]
(a) Find the speed of the block at point B immediately after it strikes the second incline.
(b) Find the speed of the block at point C, immediately before it leaves the second incline.
(c) If collision between the block and the incline is completely elastic, then find the vertical (upward) component of the velocity of the block at point B, immediately after it strikes the second incline.

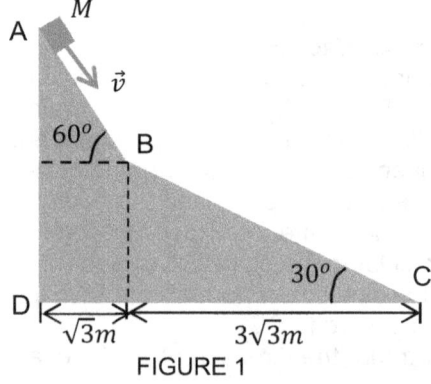

FIGURE 1

APPROACH In parts (a) and (b) the collision is completely inelastic where as in part (c) it is completely elastic collision, so we follow the similar procedure as discussed in above article.

SOLUTION From geometry $\tan 60° = \frac{h_1}{\sqrt{3}m} \Rightarrow h_1 = 3\,m$,

$\tan 30° = \frac{h_2}{3\sqrt{3}m} \Rightarrow h_2 = 3m$.

If v_B is the velocity of block at B, just before collision, then by conservation of mechanical energy between point A and point B, we have
$$Mgh_1 = \frac{1}{2}Mv_B^2.$$
$$\Rightarrow \quad v_B = \sqrt{2gh_1} = \sqrt{2 \times 10 \times 3} = \sqrt{60}\text{ m/s, parallel to AB.}$$

The linear momentum just before the collision is Mv_B parallel to AB. Resolve the linear momentum in directions parallel and perpendicular to BC to get
$$p_\parallel = Mv_B \cos 30° = M\sqrt{45},$$
$$p_\perp = Mv_B \sin 30° = M\sqrt{15}.$$

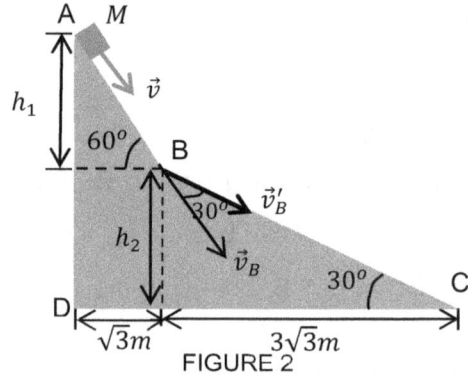

FIGURE 2

The collision at B is completely inelastic. The component of linear momentum parallel to BC is conserved but component perpendicular to BC becomes zero. Thus,
$$p_\parallel = M\sqrt{45} = Mv_B'$$
which gives the velocity of the block immediately after the collision, $v'_B = \sqrt{45}$ m/s.

(b) The conservation of energy between the point B and the point C gives
$$\frac{1}{2}Mv'^2_{v_B} + Mgh_2 = \frac{1}{2}Mv_C^2$$
Substitute the values to get the velocity of the block at the point C i.e.,
$$\frac{1}{2} \times 45 + 30 = \frac{1}{2}v_C^2 \quad \Rightarrow \quad v_C = 105 \text{ m/s}.$$

(c) In part (a), we have already calculated, the components of linear momentum just before the collision parallel and perpendicular to BC. These are

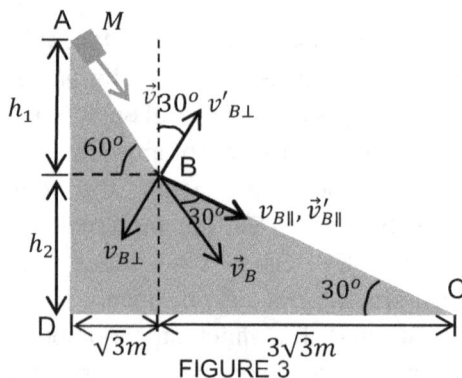

FIGURE 3

$p_\| = M v_B \cos 30^\circ = M\sqrt{45}$; $p_\perp = M v_B \sin 30^\circ = M\sqrt{15}$.

As the collision at B is completely elastic, the linear momentum of the block parallel to BC remains conserved and the magnitude of the linear momentum perpendicular to BC remains same but its direction gets reversed. i.e.,
$M\sqrt{45} = M v'_{B\|}$ or $v'_{B\|} = \sqrt{45}$ m/s
and $M v_{B\perp} = M v'_{B\perp}$ (note carefully that only magnitudes are equal but directions are opposite)
$M\sqrt{15} = M v'_{B\perp}$ or $v'_{B\perp} = \sqrt{15}$ m/s
The vertical component of the velocity just after the collision is

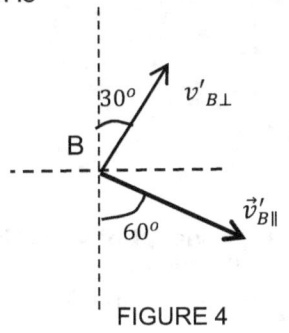

FIGURE 4

$v'_{B,vert} = v'_{B\perp} \cos 30^\circ - v'_{B\|} \cos 60^\circ$
$= \sqrt{15} \times \frac{\sqrt{3}}{2} - \sqrt{45} \times \frac{1}{2} = 0$

7. CHECKPOINT 2

1. •••Two 22.7-kg ice sleds are placed a short distance apart, one directly behind the other, as shown in Fig. P1. A 3.63-kg cat, standing on one sled, jumps across to the other and immediately back to the first. Both jumps are made at a speed of 3.05 m/s relative to the sled the cat is standing on when the jump is made. Find the final speeds of the two sleds.

FIGURE P1

2. ••Two blocks of masses m and 3m are placed on a frictionless, horizontal surface. A light spring is attached to the more massive block, and the blocks are pushed together with the spring between them (Fig. P2). A cord initially holding the blocks together is burned; after that happens, the block of mass 3m moves to the right with a speed of 2.00 m/s. (a) What is the velocity of the block of mass m? (b) Find the system's original elastic potential energy, taking $m = 0.350$ kg. (c) Is the original energy in the spring or in the cord? (d) Explain your answer to part (c). (e) Is the momentum of the system conserved in the bursting-apart process? Explain how that is possible considering (f) there are large forces acting and (g) there is no motion beforehand and plenty of motion afterward?

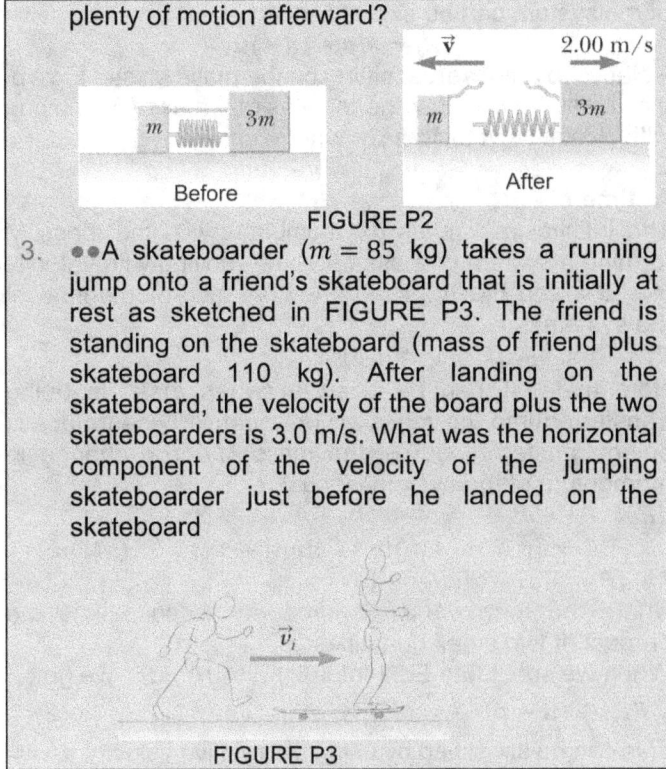

FIGURE P2

3. ••A skateboarder ($m = 85$ kg) takes a running jump onto a friend's skateboard that is initially at rest as sketched in FIGURE P3. The friend is standing on the skateboard (mass of friend plus skateboard 110 kg). After landing on the skateboard, the velocity of the board plus the two skateboarders is 3.0 m/s. What was the horizontal component of the velocity of the jumping skateboarder just before he landed on the skateboard

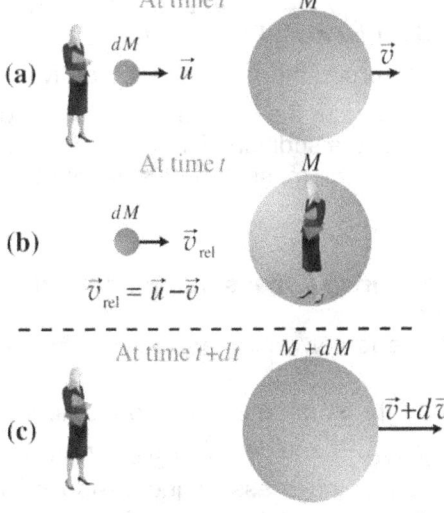

FIGURE P3

8. SYSTEM OF VARIABLE MASS

8.1. SYSTEMS OF INCREASING MASS

For the general treatment of systems of increasing mass, we use Fig.1 that depicts the following:

FIGURE. 1. (a) At time t, the differential mass dM is about to combine with the mass M. (b) The velocity of dM as seen by an observer on M at the same time t. (c) At time $t + dt$, the mass dM has combined with M

At time t
We have a system consisting of mass M moving with velocity v and momentum Mv. Also, we have an infinitesimal mass dM moving with velocity u and momentum $dM\,u$, see Fig.1a. The initial total momentum

of the system can be expressed as:
$$P_i = Mv + (dM)u$$
Relative to an observer sitting on the mass M, see Fig.1b, the observer will view the infinitesimal mass dM moving with a relative velocity v_{rel} where:
$$v_{rel} = u - v$$

At time $t + dt$
The infinitesimal mass dM combines with the mass M forming a system of mass $M + dM$ moving with velocity $v + dv$, see Fig.1c. Then, the final total momentum of the system is:
$$P_f = (M + dM)(v + dv)$$
Note that dM can be positive (when mass is being transferred into the mass M) or negative (when mass is being transferred out of the mass M). The change in momentum of the system is thus:
$$dP = P_f - P_i$$
$$= [(M + dM)(v + dv)] - [Mv + (dM)u]$$
$$\Rightarrow dP = Mdv - dM(u - v) \quad \ldots (1)$$
where the term $dM\,dv$ is dropped because it is the product of two small quantities.
When we substitute Eq.1 into $F_{ext} = dP/dt$, we get:
$$\sum F_{ext} + (u - v)\frac{dM}{dt} = M\frac{dv}{dt} \quad \ldots (2)$$
This can be simplified by using the relative velocity $v_{rel} = u - v$, such as:
$$\sum F_{ext} + v_{rel}\frac{dM}{dt} = M\frac{dv}{dt}$$
or $\quad \sum F_{ext} + F_{th} = M\frac{dv}{dt}$
$$\Rightarrow F_{net} = M\frac{dv}{dt} \quad \ldots (3)$$
The right-hand side of this equation, Mdv/dt, refers to the mass times the acceleration. The first term on the left-hand side of the equation, $\sum F_{ext}$, refers to the external force on the mass M. The second term on the left-hand side, $F_{th}\left(= v_{rel}\frac{dM}{dt}\right)$, refers to the force exerted on M, in terms of the rate at which the momentum is being transferred into M (due to the addition of mass). This term is also called the thrust on the mass M and is denoted by F_{th}.
$$F_{th} = v_{rel}\frac{dM}{dt}$$

Problems related to variable mass can be solved in following four steps
1. Make a list of all the forces acting on the main mass and apply them on it.
2. Apply an additional thrust force $\vec{F}_{th}$ on the mass, the magnitude of which is $v_{rel}\frac{dM}{dt}$ and direction is given by the direction of $\vec{v}_{rel}$ in case the mass is increasing and otherwise the direction of $-\vec{v}_{rel}$ if it is decreasing.
3. Find net force on the mass and apply $\vec{F}_{net} = M\frac{d\vec{v}}{dt}$ (m = mass at the particular instant)
4. Integrate it with proper limits to find velocity at any time t.

Note: Problems of one-dimensional motion can be solved in easier manner just by assigning positive and negative signs to all vector quantities. Here are few examples in support of the above theory.

EXAMPLE 20. FREIGHT CAR AND HOPPER A flat car of mass m_0 starts moving to the right due to a constant horizontal external force F_{ext}. Sand spills on the flat car from a stationary hopper. The rate of loading is constant and equal to μ kg/s. Find the time dependence of the velocity and the acceleration of the flat car in the process of loading. The friction is negligibly small.

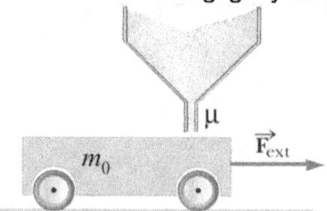

APPROACH 1: By applying concept of variable mass
SOLUTION Initial velocity of the flat car is zero. Let v be its velocity at time t and m its mass at that instant. Then

At $t = 0$, $v = 0$ and $m = m_0$ at $t = t$, $v = v$ and $m = m_0 + \mu t$
Here, $v_r = v$ \qquad (backward)
$\frac{dm}{dt} = \mu$
$F_{th} = v_r \frac{dm}{dt} = \mu v$ \qquad (backward)
Net force on the flat car at time t is $F_{net} = F - F_{th}$
$$m\frac{dv}{dt} = F_{ext} - \mu v \quad \ldots (1)$$
$$\Rightarrow (m_0 + \mu t)\frac{dv}{dt} = F_{ext} - \mu v$$
or $\int_0^v \frac{dv}{F_{ext} - \mu v} = \int_0^t \frac{dt}{m_0 + \mu t}$
$$\Rightarrow -\frac{1}{\mu}[\ln(F - \mu v)]_0^v = \frac{1}{\mu}[\ln(m_0 + \mu t)]_0^t$$
$$\Rightarrow -\frac{1}{\mu}\left[\ln\left(\frac{F_{ext} - \mu v}{F}\right)\right]_0^v = \frac{1}{\mu}\left[\ln\left(\frac{m_0 + \mu t}{m_0}\right)\right]_0^t$$
$$\Rightarrow \ln\left(\frac{F_{ext}}{F - \mu v}\right) = \ln\left(\frac{m_0 + \mu t}{m_0}\right)$$
$$\Rightarrow \frac{F_{ext}}{F - \mu v} = \frac{m_0 + \mu t}{m_0}$$
$$\Rightarrow F_{ext}m_0 = F_{ext}m_0 - \mu m_0 v + \mu F_{ext}t - \mu^2 vt$$
$$\Rightarrow 0 = -\mu m_0 v + \mu F_{ext}t - \mu^2 vt$$
$$\Rightarrow v(m_0 + \mu t) = F_{ext}t$$
$$\Rightarrow v = \frac{F_{ext}t}{m_0 + \mu t} \qquad \text{Ans.}$$
Acceleration of flat car at time t is

LINEAR MOMENTUM, IMPULSE AND COLLISIONS

$$a = \frac{dv}{dt} = \frac{F_{ext} + \mu v}{m} = \frac{F_{ext} + \mu \frac{Ft}{m_0 - \mu t}}{m_0 + \mu t} = \frac{F_{ext} m_0}{(m_0 + \mu t)^2}$$

APPROACH 1: By using variable mass and force relationship

Equations to be used are

$$\Sigma F_{ext} + (u - v)\frac{dM}{dt} = M\frac{dv}{dt} \qquad \ldots (1)$$

This can be simplified by using the relative velocity $v_{rel} = u - v$, such as:

$$\Sigma F_{ext} + v_{rel}\frac{dM}{dt} = M\frac{dv}{dt} \Rightarrow F_{net} = M\frac{dv}{dt} \qquad \ldots (2)$$

and thrust

$$F_{th} = v_{rel}\frac{dM}{dt} \qquad \ldots (3)$$

From (1)

$$F_{ext} + (0 - v)\mu = m\frac{dv}{dt} \quad \text{or} \quad F_{ext} - \mu v = m\frac{dv}{dt}$$

Now we can solve as above.

EXAMPLE 21. Figure shows a stationary hopper that drops sand at a rate $dM/dt = 5\ kg/s$ onto a conveyer belt. The belt is supported by frictionless rollers and moves at a constant speed $v = 1.2\ m/s$ under the action of a constant external force $\vec{F}_{ext}$. (a) Find the value of the external force $\vec{F}_{ext}$ that is needed to keep the belt moving with a constant speed. (b) Find the power delivered by the external force $\vec{F}_{ext}$. (c) Find the rate of the kinetic energy acquired by the falling sand due to the change in its horizontal motion.

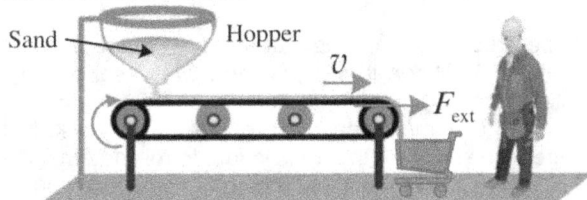

APPROACH We solve this problem by directly applying the relation-

$$\Sigma F_{ext} + (u - v)\frac{dM}{dt} = M\frac{dv}{dt} \qquad \ldots (1)$$

SOLUTION (a) Since, the hopper is stationary, therefore $u = 0$. We also take $dv/dt = 0$ because the belt is moving with constant speed. Thus, from Eq. 1, we get-

$$\Sigma F_{ext} + (u - v)\frac{dM}{dt} = M\frac{dv}{dt} \Rightarrow F_{ext} + (0 - v)\frac{dM}{dt} = 0$$

$$\Rightarrow F_{ext} = v\frac{dM}{dt}$$

$$\therefore F_{ext} = (1.2\ m/s)(5\ kg/s) = 6\ N$$

The only horizontal force on the sand is the friction of the belt f_s. Thus,

$$f_s = F_{ext}.$$

(b) The power delivered by $\vec{F}_{ext}$ is work done by this force in 1s. Thus:

$$P = \frac{dW}{dt} = \vec{F}_{ext} \cdot \vec{v} = F_{ext}\, v = v^2 \frac{dM}{dt} = (1.2\ m/s)^2 (5\ kg/s)$$
$$= 7.2\ W$$

This work per unit time is the power output required by the motor.

(c) The rate of the kinetic energy acquired by the falling sand is:

$$\frac{dK}{dt} = \frac{d}{dt}\left(\frac{1}{2}Mv^2\right) = \frac{1}{2}\frac{dM}{dt}v^2 = \frac{1}{2}(5\ kg/s)(1.2\ m/s)^2 = 3.6\ W$$

This is only half the power delivered by $\vec{F}_{ext}$. The other half goes into thermal energy produced by friction between the sand and the belt.

EXAMPLE 22. LEAKY FREIGHT CAR A cart loaded with sand moves along a horizontal floor due to a constant force F coinciding in direction with the cart's velocity vector. In the process sand spills through a hole in the bottom with a constant rate μ kg/s. Find the acceleration and velocity of the cart at the moment t, if at the initial moment t = 0 the cart with loaded sand had the mass m_0 and its velocity was equal to zero. Friction is to be neglected.

APPROACH. If at any instant the velocity of the cart is v, then we have

$$F_{ext} + (u - v)\frac{dm}{dt} = m\frac{dv}{dt} \qquad \ldots (1)$$

As the sand spills through the hole in the bottom of the cart, so it will always acquire the same velocity as that of the cart at that moment i.e., $u = v$, and the external force is $F_{ext} = F$, therefore from (1), we have

$$F = m\frac{dv}{dt}$$

Here, mass at time 't' will be given by, $m = m_0 - \mu t$

$$\therefore \quad (m_0 - \mu t)\frac{dv}{dt} = F \qquad \ldots (2)$$

SOLUTION From equation (2), we can write

$$\int_0^v dv = \int_0^t \frac{F\, dt}{m_0 - \mu t} \quad \text{or} \quad v = -\frac{F}{\mu}[\ln(m_0 - \mu t)]_0^t$$

or $v = -\frac{F}{\mu}\left[\ln\left(\frac{m_0 - \mu t}{m_0}\right)\right] = \frac{F}{\mu}\left[\ln\left(\frac{m_0}{m_0 - \mu t}\right)\right]$ **Ans.**

Acceleration $a = \frac{dv}{dt} = \frac{F}{M} = \frac{F}{m_0 - \mu t}$ **Ans.**

EXAMPLE 23. LEAKY FREIGHT CAR AND HOPPER
Sand falls from a stationary hopper into a freight car which is moving with uniform velocity v_0. The sand falls at the rate μ into the car and the same freight car is leaking sand at the rate μ, thus keeping its mass constant (see adjoining FIGURE). How much force is needed to keep the freight car moving at the speed v_0?

SOLUTION consider the freight car, the sand inside it and small amount of sand which is going to fall from hopper onto the car in time Δt as one system. This small amount of sand is in mid air, falling freely right now and does not have any horizontal component of velocity.

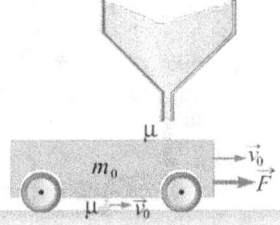

Thus, linear momentum of the system in horizontal direction at time t is

$$P_i = m_0 v_0 + \mu \Delta t (0)$$

Once the small amount, $\mu \Delta t$ of the sand joins the

remaining sand in the car, it will have same velocity as that of the car due to impulsive friction. Also, during the same time, small amount of the sand leaves the car having mass, $\mu \Delta t$. However, the velocity of the sand after leaving the freight car is same as its initial velocity, and hence its momentum does not change. Thus, the momentum of the system in horizontal direction at time $t + dt$ is:

$$P_f = [(m_0 - \mu \Delta t)v_0 + \mu \Delta t v_0] + \mu \Delta t v_0$$

Using, $F = \dfrac{\Delta P}{\Delta t} = \dfrac{P_f - P_i}{\Delta t}$

or $\quad F \Delta t = P_f - P_i = \mu \Delta t v_0$

or $\quad F = \mu v_0$ \hfill Ans.

☞ We encourage you to solve the above problem by direct method.

9. CHECKPOINT 3

1. ●●●Sand falls from a stationary hopper into a freight car which is moving with uniform velocity v_0. The sand falls at the rate μ. How much force is needed to keep the car moving at the speed v_0?

2. ●●•If a freight car is leaking sand at the rate μ, what force is needed to keep the freight car moving uniformly with speed v_0.

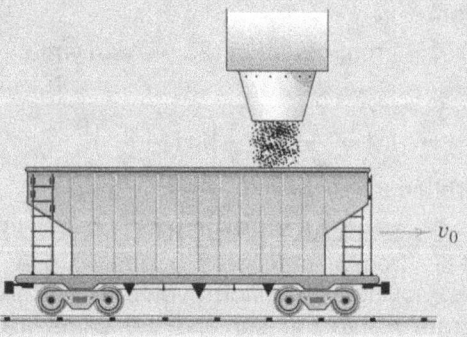

FIGURE P2

3. ●●●A coal car with an empty mass of 25 Mg is moving freely with a speed of 1.2 m/s under a hopper which opens and releases coal into the moving car at the constant rate of 4 Mg per second. Determine the distance x moved by the car during the time that 32 Mg of coal are deposited in the car. Neglect any frictional resistance to rolling along the horizontal track. [$1 Mg$(megagram)$= 10^3$ kg]

4. ●●●Sand is released from the hopper H with negligible velocity and then falls a distance h to the conveyor belt. The mass flow rate from the hopper is μ. Develop an expression for the steady-state belt speed v for the case $h = 0$. Assume that the sand quickly acquires the belt velocity with no rebound, and neglect friction at the pulleys A and B.

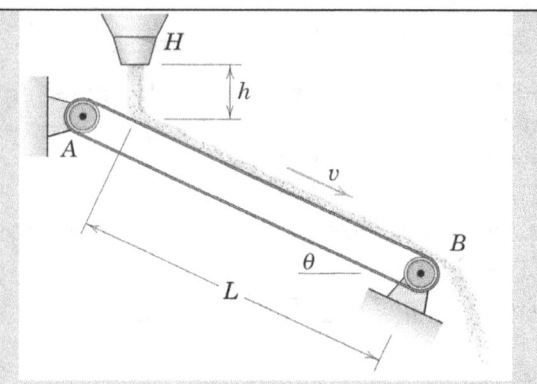

FIGURE P4

5. ●●●Repeat the previous problem, but now let $h \neq 0$. Then evaluate your expression for the conditions $h = 2$ m, $L = 10$ m, and $\theta = 25°$.

6. ●●●The end of a pile of loose-link chain of mass ρ per unit length is being pulled horizontally along the surface by a constant force F. If the coefficient of kinetic friction between the chain and the surface is μ_k, determine the acceleration a of the chain in terms of x and velocity v.

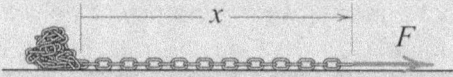

FIGURE P6

7. ●●●The truck has a mass of 50 Mg when empty (FIGURE P6). When it is unloading 5 m^3 of sand at a constant rate of 0.8 m^3/s, the sand flows out the back at a speed of 7 m/s, measured relative to the truck, in the direction shown. If the truck is free to roll, determine its initial acceleration just as the sand begins to fall out. Neglect the mass of the wheels and any frictional resistance to motion. The density of sand is $\rho_s = 1520$ kg/m^3. [$1Mg$(megagram)$= 10^3 kg$]

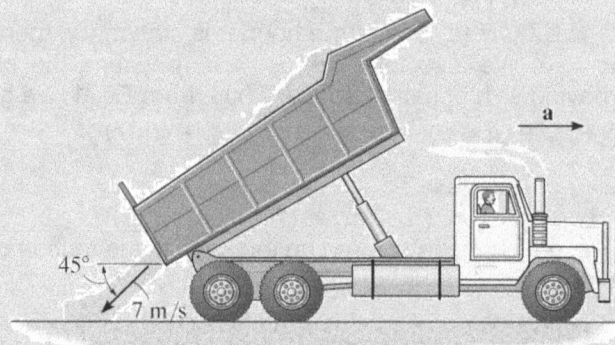

FIGURE P7

8. ●●●The truck has a mass m_0 and is used to tow the smooth chain having a total length l and a mass per unit of length ρ [Fig P7]. If the chain is originally piled up, determine the tractive force F that must be supplied by the rear wheels of the truck necessary to maintain a constant speed v while the chain is being drawn out.

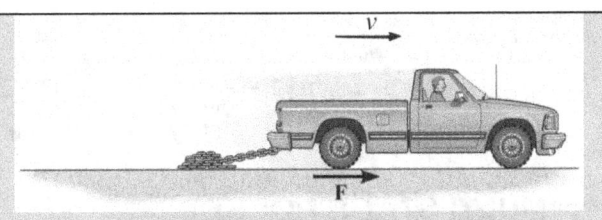

FIGURE P8

9. •••A coil of heavy flexible cable with a total length of 100 m and a mass of 1.2 kg /m is to be laid along a straight horizontal line. The end is secured to a post at A, and the cable peels off the coil and emerges through the horizontal opening in the cart as shown. The cart and drum together have a mass of 40 kg. If the cart is moving to the right with a velocity of 2 m /s when 30 m of cable remain in the drum and the tension in the rope at the post is 2.4 N, determine the force F required to give the cart and drum an acceleration of 0.3 m/s^2. Neglect all friction.

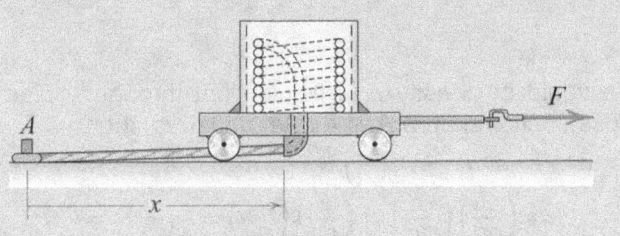

FIGURE P9

10. •••The mass m of a raindrop increases as it picks up moisture during its vertical descent through still air. If the air resistance to motion of the drop is R and its downward velocity is v, write the equation of motion for the drop and show that the relation $\sum F = \frac{d}{dt}(mv)$ is obeyed as a special case of the variable-mass equation.

11. •••Water falls without splashing at a rate of 0.250 L/s from a height of 2.60 m into a 0.750-kg bucket on a scale. If the bucket is originally empty, what does the scale read in newtons 3.00 s after water starts to accumulate in it?

10. SYSTEMS OF DECREASING MASS; ROCKET PROPULSION

Now we treat systems with decreasing mass by considering the case of rocket propulsion as shown in FIGURE 1.

At time t
We have a system boundary consisting of a rocket of mass M moving with velocity v and momentum Mv, see Fig.1a. The initial total momentum of the system can be expressed as:
$$P_i = Mv \quad \ldots (1)$$

At time t + dt
We have a system boundary consisting of a rocket of mass $M - dM$ moving with velocity $v + dv$ and an ejected exhaust of mass dM moving with velocity u, see Fig.1b. The final total momentum of the system boundary is:
$$P_f = (M - dM)(v + dv) + (dM)u \quad \ldots (2)$$

The relative velocity of exhaust in forward direction with respect to observer sitting on the rocket (Fig 1c) $= u - (v + dv)$.

But in problems, generally this velocity is given in backward direction, so we convert it in backward direction by just multiplying with $-ve$ sign, we get
The relative velocity of exhaust in backward direction with respect to the observer sitting in the rocket
$$v_{rel} = -[u - (v + dv)] = (v + dv) - u$$

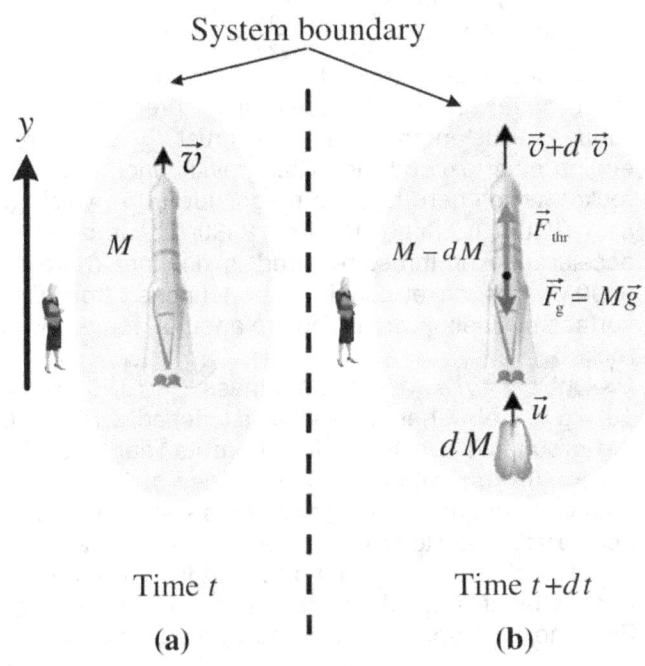

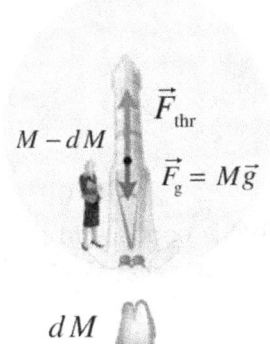

FIGURE 1 (a) At time t, the rocket has a mass M. (b) At time $t + dt$, the mass of the exhaust dM has been ejected from M. (c) The velocity of the exhaust dM as seen by an observer on the rocket at time $t + dt$

or $u = (v+dv) - v_{rel}$
Therefore, (2) becomes $P_f = (M-dM)(v+dv) + (dM)[(v+dv) - v_{rel}]$
$P_f = Mv + Mdv - (dM)v - dMdv$
$\qquad + (dM)v + dMdv - (dM)v_{rel}$
$P_f = Mv + Mdv - (dM)v + (dM)v - (dM)v_{rel}$ ... (3)
Change in linear momentum
$dP = P_f - P_i$
$\quad = Mv + Mdv - (dM)v + (dM)v - (dM)v_{rel} - Mv$
$\Rightarrow dP = Mdv - (dM)v_{rel}$
$\Rightarrow \dfrac{dP}{dt} = M\dfrac{dv}{dt} - v_{rel}\dfrac{dM}{dt}$ ... (4)

Now $\dfrac{dP}{dt} = F_{ext}$, external force acting on Rocket and the term $v_{rel}\dfrac{dM}{dt}$ refers to the force exerted on M in terms of the rate at which the momentum is being transferred out of M (due to the ejection of mass). For rockets, this term is positive since dM/dt is negative and v_{rel} is negative (opposite to). This term is called the thrust, F_{th}, and represents the force exerted on the rocket by the ejected gasses. Thrust is defined as follows:

$F_{th} = v_{rel}\dfrac{dM}{dt}$, therefore

$F_{ext} = M\dfrac{dv}{dt} - F_{th}$ ... (5)

Where, $F_{th} \Rightarrow$ thrust
$\Rightarrow F_{ext} + F_{th} = M\dfrac{dv}{dt} \Rightarrow F_{net} = M\dfrac{dv}{dt}$,
Here
$F_{net} = F_{ext} + F_{th}$
= net force on rocket
In one-dimensional vertical motion under a constant gravitational force, where
$F_{ext} = -Mg$, we can find the speed of the rocket at any time t, by rewriting Eq.4. as:
$-Mg + v_{rel}\dfrac{dM}{dt} = M\dfrac{dv}{dt}$
$\Rightarrow M\dfrac{dv}{dt} = -Mg + v_{rel}\dfrac{dM}{dt}$
$\Rightarrow \dfrac{dv}{dt} = -g + v_{rel}\dfrac{1}{M}\dfrac{dM}{dt}$
$\Rightarrow dv = -gdt + v_{rel}\dfrac{1}{M}dM$
Integrating both sides, we get
$\int_{v_0}^{v} dv = -g\int_0^t dt + v_{rel}\int_{M_0}^{M}\dfrac{1}{M}dM$
$\Rightarrow v - v_0 = -gt + v_{rel}\ln\dfrac{M}{M_0}$
$\Rightarrow v = v_0 - gt + v_{rel}\ln\dfrac{M}{M_0}$ ... (6)
Note that v_{rel} is negative because it is opposite to the rocket's motion and $\ln\dfrac{M}{M_0}$ is also negative because $M_0 > M$.
If rate of burning of fuel is 'μ', then
$\mu = -\dfrac{dM}{dt}$
(Negative sign shows that remaining mass decreases with time.)
$\Rightarrow \int_{M_0}^{M} dM = -\mu\int_0^t dt \Rightarrow M = M_0 - \mu t$
Therefore equation (6) can also be written as:

$v = v_0 - gt + v_{rel}\ln\dfrac{M_0 - \mu t}{M_0}$
or $\quad v = v_0 - gt + v_{rel}\ln\left(1 - \dfrac{\mu t}{M_0}\right)$
The above expression gives the velocity of rocket at time 't'.

10.1. MULTI STAGE ROCKET

A **multistage** (or **multi-stage**) **rocket** is a rocket that uses two or more *stages*, each of which contains its own engines and propellant.

Note: Above method is also applicable for liquid flowing out through a pipe.

If a liquid of density ρ is flowing out through a pipe of cross-sectional area A at a rate of Q m³/s, then

Thrust, $F_{th} = v\dfrac{dM}{dt} = \left(\dfrac{dl}{dt}\right)\dfrac{\rho dV}{dt}$
$= \left(\dfrac{Adl}{Adt}\right)\left(\dfrac{\rho dV}{dt}\right) = \left(\dfrac{dV}{Adt}\right)\left(\dfrac{\rho dV}{dt}\right) = \dfrac{\rho}{A}\left(\dfrac{dV}{dt}\right)^2 = \dfrac{\rho}{A}Q^2$

EXAMPLE 24. A rocket whose initial mass M_0 is 850 kg consumes fuel at the rate $R = 2.3$ kg/sec. The speed v_{rel} of the exhaust gases relative to the rocket engine is 2800 m/sec.
(a) What thrust does the rocket engine provide?
(b) What is the initial acceleration of the rocket?

SOLUTION (a) Thrust, $F_{th} = v_{rel}\dfrac{dM}{dt} = (2800 m/s)(2.3 kg/s) = 6440 N$

(b) Acceleration, $a = \dfrac{F_{th}}{M_0} = \dfrac{6440N}{850kg} = 7.6 m/s^2$

Conclude: To be launched from Earth's surface, a rocket must have an initial acceleration greater than $g = 9.8 m/s^2$. Put another way, the thrust F_{th} of the rocket engine must exceed the initial gravitational force on the rocket, which here has the magnitude $M_0 g$, which gives us $(850kg)(9.8m/s^2)$, or $8330N$. Because the acceleration or thrust required is not met (here $F_{th} = 6400N$), our rocket could not be launched from Earth's surface by itself; it would require another, more powerful, rocket.

EXAMPLE 25. A rocket has a mass 2×10^4 kg of which 10^4 kg is fuel. When the rocket is lunched vertically from the ground, it consumes fuel from its rear at a rate of 1.5×10^3 kg/s with an exhaust speed of 2.5×10^3 m/s relative to the rocket. Neglect air resistance and take the acceleration due to gravity to be $= 9.8 m/s^2$. (a) Find the thrust on the rocket. (b) Find the net force on the rocket, once when it is full of fuel and once when it is empty. (c) Find the final speed of the rocket when the fuel burns completely.

SOLUTION (a) Since the motion is in one dimension and we can take upward as positive, then v_{rel} is negative because it is downward and dM/dt is negative because the rocket's mass is decreasing. Therefore, the thrust is:
$$F_{th} = v_{rel}\frac{dM}{dt}$$
$$= \left(-2.5 \times \frac{10^3 m}{s}\right)(-1.5 \times 10^3 kg/s) = 3.75 \times 10^6 N$$

(b) Initially the net force on the rocket is:
$$F_{net} = F_{th} - M_0 g$$
$$= (3.75 \times 10^6 N) - (2 \times 10^4 kg)(9.8 m/s^2)$$
$$= 3.554 \times 10^6 N$$

The net force just before the rocket is out of fuel is:
$$F_{net} = F_{th} - M_0 g$$
$$= (3.75 \times 10^6 N) - (1 \times 10^4 kg)(9.8 m/s^2)$$
$$= 3.652 \times 10^6 N$$

(c) The time required to reach fuel burnout is the time needed to use all the fuel (10^4 kg) at rate of 1.5×10^3 kg/s. Thus:
$$t = \frac{10^4 kg}{1.5 \times 10^3 kg/s} = 6.67 s$$

By taking $v_0 = 0$ and using Eq.6, we find that:
$$v - v_0 = -gt + v_{rel} \ln\frac{M}{M_0}$$
$$\Rightarrow v = -\left(9.8\frac{m}{s^2}\right)(6.67 s)$$
$$+ \left(-2.5 \times \frac{10^3 m}{s}\right) \times \ln\left(\frac{1 \times 10^4 kg}{2 \times 10^4 kg}\right) = 1667.5 m/s$$

EXAMPLE 26. In the FIGURE 1, the block A has mass m_1 (constant) and B has initial mass m_0. B is filled with sand which is thrown out by some internal mechanism at constant rate μ kg/sec at velocity v relative to B in downward direction. Assuming $m_1 > m_0$, light string and pulley, no friction in pulley and motion in vertical plane find the acceleration of A.

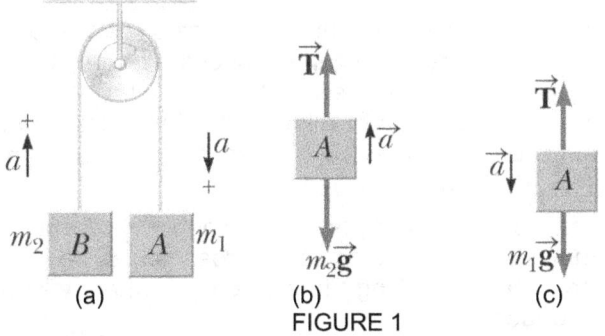

FIGURE 1

APPROACH As block B is decreasing mass system, and its initial mass (m_0) is less than that of A, so its acceleration will be in upward direction and it is time dependent. Suppose at any instant 't' its upward acceleration is $\frac{dv}{dt} = a$, so for it we have to apply the equation
$$F_{ext} + v_{rel}\frac{dm}{dt} = m\frac{dv}{dt}$$
here, $F_{ext} = T - m_2 g$, $v_{rel} = -v$ and $\frac{dm}{dt} = -\mu$ (negative sign is used for decreasing mass) and $m_2 = m_0 - \mu t$ is the mass at any instant t.
$$\therefore (T - m_2 g) - v(-\mu) = m_2 a$$
or $(T - m_2 g) + \mu v = m_2 a$ (for B) ...(1)

Block A has constant mass m_1, and it is attached with B by inextensible light string so at time t it will also have acceleation a in downward direction. Now by Newton's second law of motion for A, we have
$$m_1 g - T = m_1 a \quad \text{(for A)} \quad ...(2)$$

SOLUTION Adding equations, (1) and (2), we get
$$a = \frac{(m_1 - m_2)g + \mu v}{m_1 + m_2}, \quad m_2 = m_0 - \mu t$$

11. CHECKPOINT 4

1. ●●At the instant of vertical launch, the rocket expels exhaust at the rate of 220 kg/s with an exhaust velocity of 820 m/s. If the initial vertical acceleration is 6.80 m/s², calculate the total mass of the rocket and fuel at launch.

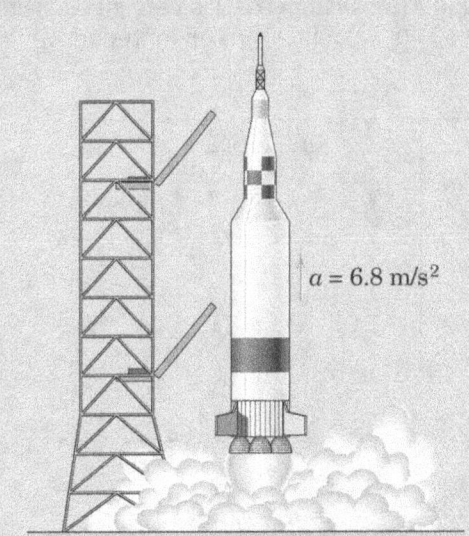

FIGURE P1

2. ●●A small rocket of initial mass m_0 is fired vertically upward near the surface of the earth (g constant). If air resistance is neglected, determine the manner in which the mass m of the rocket must vary as a function of the time t after launching in order that the rocket may have a constant vertical acceleration a, with a constant relative velocity u of the escaping gases with respect to the nozzle.

3. ●●A garden hose is held as shown in FIGURE P3. The hose is originally full of motionless water. What additional force is necessary to hold the nozzle stationary after the water flow is turned on if the discharge rate is 0.600 kg/s with a speed of 25.0 m/s?

FIGURE P3

12. CHAIN RELATED PROBLEMS

To understand chain related problems, we are giving some examples with correct approach and solution steps.

EXAMPLE 27. A uniform chain $A'B'$ of length $2l$ having mass λ per unit length is hanging from ceiling by two light, inextensible threads AA' and BB' of equal length as shown in Fig.1(a). Distance AB is very small. Thread BB' is burnt at $t = 0$, calculate tension in thread AA' at time t.

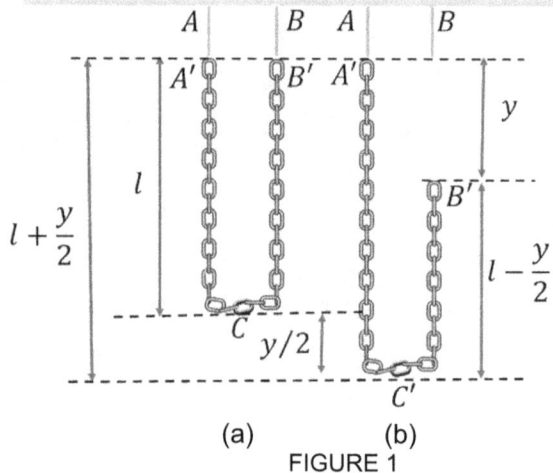

(a) (b)
FIGURE 1

APPROACH When the thread BB' is burnt, end B' of chain starts moving under gravity while A' is at rest. Hence, length of left side vertical part of chain increases. Now, in this situation the tension in thread AA' will be due to two factors:
(1) due to weight of left side vertical part of chain, and
(2) due to rate of change of momentum of the chain when its links transfers from right to left part.
So, we will try to find the mass on the left side as a function of time and apply impulse momentum equation for continuous process to the chain on the left side.

SOLUTION Consider downward motion of end B' as shown in FIGURE 1b
Initial relative velocity, $u = 0$
Acceleration, $a' = g$
Using equation of motion, $y = ut + \frac{1}{2}a't^2$
We calculate relative displacement of end B' in time t as:
$y = \frac{1}{2}gt^2$ [as $u = 0$ and $a' = g$]

Mass of left vertical part of chain at time t is, $m = \left(l + \frac{y}{2}\right)\lambda$
Putting the value of y, we get, $m = \lambda l + \frac{1}{4}\lambda g t^2$
Let rate of increase of mass in left part be, $\mu = \frac{dm}{dt}$
Using the value of m from above, we get, $\mu = \frac{1}{2}\lambda g t$
Change in velocity of a link when it transfers from right to left vertical part is $\Delta v = gt$
Therefore, rate of change of momentum is
$$\frac{dp}{dt} = \mu \Delta v = \frac{1}{2}\lambda g^2 t^2$$
Weight of left vertical part, $mg = \left[\lambda l + \frac{1}{4}\lambda t^2\right]g$
Therefore, required tension, $T = mg + \frac{dp}{dt}$
Putting values from above equation, we get
$$T = \lambda l g + \frac{1}{2}\lambda g^2 t^2 \qquad \text{Ans.}$$

DISCUSSION The tension is more than the weight of the left side as every part joining the chain from right part applies impulse on it in vertically downward direction as chain applies force on the element to stop its vertically downward motion.

12.1. FORCE EXERTED BY FREE FALL OF CHAIN ON A SURFACE

Force F exerted by chain consists of two components
(a) weight of the fallen portion of the chain,
(b) thrust of the falling part of chain.
When a chain is released all its links will be in a state of free fall so the chain always remains vertical but without tension. After time t the tip link will fall a distance $\frac{1}{2}gt^2$.
Step 1: Find the fall out length on the surface in time t
Step 2. Find weight of this length
Step 3: Find the normal reaction N of the weight normal to the surface
Step 4: Calculate the extra impulsive force on the surface due to change in linear momentum of chain as follows-
(i) Length fallen in time interval dt is $dy = vdt = gtdt$
(ii) Now calculate mass of this length. If total length of chain is L and its mass is M, then mass of length dx will be $dm = \frac{m}{L}dy = \frac{m}{L}gtdt$
(iii) Calculate the velocity and linear momentum components of this mass perpendicular to the surface before striking the surface and after striking the surface.
(iv) Apply the relation $F = \frac{\Delta p}{\Delta t}$ and calculate F
Step 5. Total force on surface at any time t is $F_N = N + F$
Step 6. Calculation of time of fall: $s = ut + \frac{1}{2}gt^2$
$$\Rightarrow L = (1/2)gt^2 \Rightarrow t = \sqrt{2L/g}$$
Step 6. Total impulsive force
On substituting $t = \sqrt{2L/g}$ in above relation, we get the required total force on surface
EXAMPLE 28. A chain of mass m and length L is held

vertical, such that its lower end just touches the floor. It released from rest. Find the force exerted by the chain on the table when upper end is about to hit the floor.

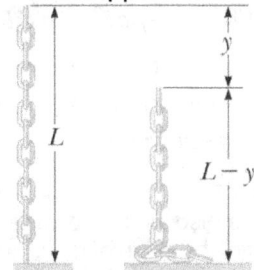

APPROACH 1. Here we fallow the steps discussed above.
SOLUTION Step 1. Find the fall out length on the surface in time t.
$$y = (1/2)gt^2$$
Step 2. Find weight of this length
Mass per unit length $\lambda = m/L$
$\therefore \quad w = \lambda yg = \frac{m}{2L}g^2t^2$

Step 3: Find the normal reaction N of the weight normal to the surface
$$N = \frac{mg^2t^2}{2L}$$
Step 4: Calculate the extra impulsive force on the surface due to change in linear momentum of chain as follows-
(i) Length fallen in time interval dt is $dy = vdt = gtdt$
(ii) Now calculate mass of this length. If total length of chain is L and its mass is m, then mass of length dx will be $\quad dm = \lambda \, dy = \frac{m}{L}dy = \frac{mgt}{L}dt$
(iii) Calculate the velocity and linear momentum components of this mass perpendicular to the surface before striking the surface and after striking the surface.
$$p_i = (dm)v_i = \left(\frac{mgt}{L}dt\right)gt = \frac{mg^2t^2}{L}dt$$
$$p_f = (dm)v_f = \left(\frac{mgt}{L}dt\right)0 = 0$$
(iv) If F is the extra impulsive force in time dt, then
$$F = \left|\frac{dp}{dt}\right| = \frac{mg^2t^2}{L}$$
The total normal force at any instant, while the chain is falling, is
$$F_N = \frac{mg^2t^2}{2L} + \frac{mg^2t^2}{L} = \frac{3mg^2t^2}{2L}$$
Time of fall: $s = ut + \frac{1}{2}gt^2$
$$\Rightarrow L = \frac{1}{2}gt^2 \Rightarrow t = \sqrt{\frac{2L}{g}}$$
The total force $= \frac{3mg^2\left(\sqrt{\frac{2L}{g}}\right)^2}{2L} = 3mg$

APPROACH 2. Force F exerted by chain consists of two components
1. F_1: weight of the fallen portion of the chain,
2. F_2: thrust of the falling part of chain, it is given by $F_2 = v_{rel}\frac{dm}{dt}$
Now, the net force exerted by chain on the floor is given by, $F = F_1 + F_2$
SOLUTION To calculate F_1 and F_2, we consider a small element of chain of length dx at height x from the surface. When this element strikes the surface, the length of chain fallen on the surface will be x and the force exerted due to its weight will be
$F_1 = mg = \lambda x g$.
Here λ is the mass per unit length of chain
The velocity of the element, at the time of striking the surface, is given by
$v^2 = u^2 + 2ax,$ here $u = 0, a = +g$, therefore
$$v = \sqrt{2gx}$$

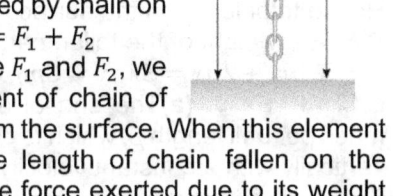

and $F_2 = v_{rel}\frac{dm}{dt}$
We have $v_{rel} = v$ and $dm = \lambda dx$
$$\therefore F_2 = v\frac{\lambda \, dx}{dt} = \lambda v^2$$
The force exerted by chain on the floor,
$$F = F_1 + F_2 = \lambda yg + \lambda v^2 = \lambda xg + \lambda\left(\sqrt{2gx}\right)^2$$
$$= \lambda xg + 2\lambda g = 3\lambda xg$$
When upper end is about to hit the floor, $x = L$
$$\therefore F = 3\lambda Lg = 3mg$$

EXAMPLE 29. A uniform chain of mass m and length L hangs by a thread and just touches the surface of a table by its lower end. Find the force exerted by the chain on the surface when half of its length has fallen on the table. Assume that the fallen part does not form heap.

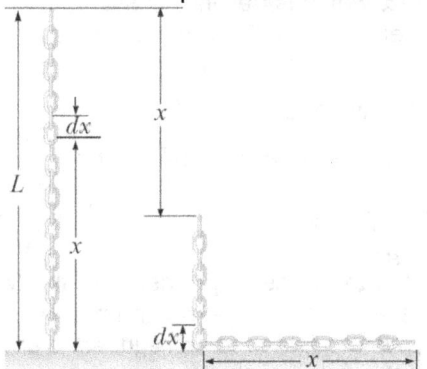

SOLUTION Consider an element of the chain of length dx at a height x from the surface. This element will hit the surface with speed $v = \sqrt{2gx}$
Mass of the element is
$dm = \lambda dx$, where $\lambda = m/L$
Hence momentum of this element before collision is $p_i = dm.v$ (downward)

Final momentum of the element is $p_f = 0$.
Therefore, change in momentum is given by
$dp = p_f - p_i = -dm \cdot v = -\lambda v dx$
$\Rightarrow \frac{dp}{dt} = -\lambda v \frac{dx}{dt} = -\lambda v^2$ as $\frac{dx}{dt} = v$
or $\quad F = \lambda v^2 = 2\lambda g x$
F is the upward force exerted by the surface on the chain. Hence total force on the surface by the chain is given by
$F' = F +$ weight of the fallen part
$= 2\lambda g x + \lambda g x = 3\lambda g x = 3mgx/L,$ when $x = L/2$

EXAMPLE 30. (a) The end of a chain of length L and mass per unit length ρ, which is piled on a platform is lifted vertically with a constant velocity v by a variable force F. Find F as a function of the height x of the end above the platform. Also find the energy lost during the lifting of the chain. (b) Also calculate force, when it is lowered with constant velocity v.

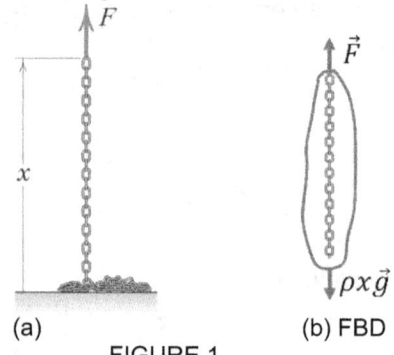

(a)   (b) FBD
FIGURE 1

(a) PART I: APPROACH This problem is similar to variable mass problems. Here the moving part of the chain of length x is gaining mass.
So, in this case we use equation,
$F_{ext} + (u-v)\frac{dm}{dt} = m\frac{dv}{dt}$ ... (1)
This equation can be applied to the moving part of the chain of length x which is gaining mass.
SOLUTION Here, $F_{ext} = F - \rho g x$, $u = 0$, $v =$ constant
i.e., $\frac{dv}{dt} = 0$
∴ from (1), we get,
$F - \rho g x + (0-v)\frac{dm}{dt} = 0$
or $F = \rho g x + v\frac{dm}{dt} = \rho g x + v(\rho v)$ $\left[\because \frac{dm}{dt} = \rho v\right]$
or $\quad F = \rho(g x + v^2)$
We now see that the force F_{ext} consists of the two parts, $\rho g x$, which is the weight of the moving part of the chain, and ρv^2, which is the added force required to change the momentum of the links on the platform from a condition at rest to a velocity v.

PART II APPROACH *Energy Loss.* Each link on the platform acquires its velocity abruptly through an impact with the link above it, which lifts it off the platform. The succession of impacts gives rise to an energy loss ΔE (negative work $-\Delta E$) so that the work-
$W_{net} = K_2 - K_1$

or $\quad \int_0^L F dx + W_g - \Delta E = K_2 - K_1$ ... (2)
here $\quad \int_0^L F dx = \int_0^L \rho(gx + v^2)dx$
$= \rho g \int_0^L x\, dx + \rho v^2 \int_0^L dx = \frac{1}{2}\rho g L^2 + \rho v^2 L$
$K_2 = \frac{1}{2}\rho L v^2$, $K_1 = 0$, $W_g = -\int_0^L \rho g x dx = -\frac{1}{2}\rho g L^2$
SOLUTION Using these values in equation (2), we get
$\frac{1}{2}\rho g L^2 + \rho v^2 L - \frac{1}{2}\rho g L^2 - \Delta E = \frac{1}{2}\rho L v^2$
or $\quad \Delta E = \frac{1}{2}\rho L v^2$ Ans.

(b) APPROACH This problem is also similar to variable mass problems. Here the moving part of the chain of length x is losing mass. When the chain is being lowered, the links which are expelled (given zero velocity) do not impart an impulse to the remaining suspended links.
So, in this case we use equation,
$F_{ext} + (u-v)\frac{dm}{dt} = m\frac{dv}{dt}$ ... (1)
This equation can be applied to the moving part of the chain of length x which is gaining mass.
SOLUTION Here, $F_{ext} = \rho g x - F$, $u = v$, $v =$ constant
i.e., $\frac{dv}{dt} = 0$
∴ from (1), we get, $\rho g x - F + (v-v)\frac{dm}{dt} = 0$
or $\quad F = \rho g x$

☞ Here we have calculated variable lifting and lowering force not thrust on the table as in previous problems.

13. CHECKPOINT 5

1. •••The upper end of the open-link chain of length L and mass ρ per unit length is released from rest with the lower end just touching the platform of the scale. Determine the expression for the force F read on the scale as a function of the distance x through which the upper end has fallen. (*Comment:* The chain acquires a free-fall velocity of $\sqrt{2gx}$ because the links on the scale exert no force on those above, which are still falling freely.

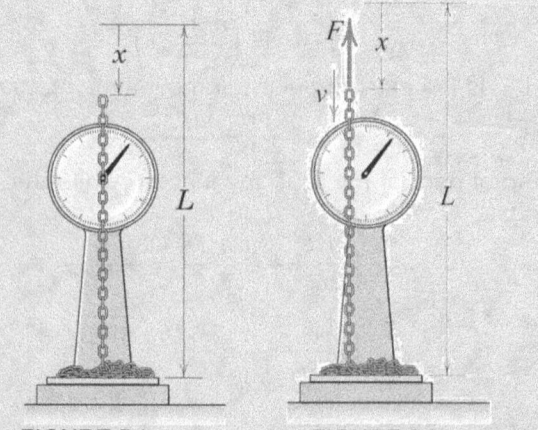

     FIGURE P1   FIGURE P2

2. •••The upper end of the open-link chain of length L and mass ρ per unit length is lowered at a constant speed v by the force F. Determine the reading N of the platform scale in terms of x.

3. ●●●The chain has a total length $L < d$ and a mass per unit length of ρ. If a portion h of the chain is suspended over the table and released, determine the velocity of its end A as a function of its position y. Neglect friction.

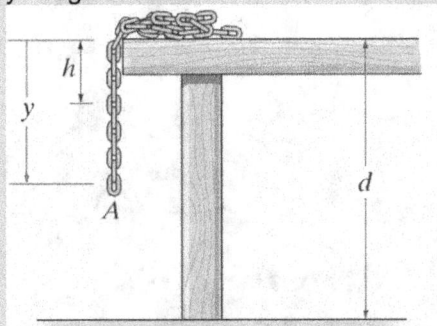

FIGURE P3

4. ●●●Determine the magnitude of force F as a function of time, which must be applied to the end of the cord at A to raise the hook H with a constant speed $v = 0.4$ m/s. Initially the chain is at rest on the ground. Neglect the mass of the cord and the hook. The chain has a mass of 2 kg/m.

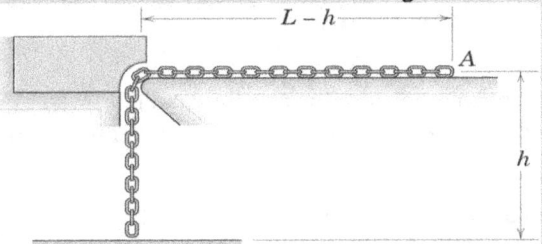

FIGURE P4

5. ●●●The free end of the flexible and inextensible rope of mass ρ per unit length and total length L is given a constant upward velocity v. Write expressions for F, the force R supporting the fixed end, and the tension T_1 in the rope at the loop in terms of x. (For the loop of negligible size, the tension is the same on both sides.)

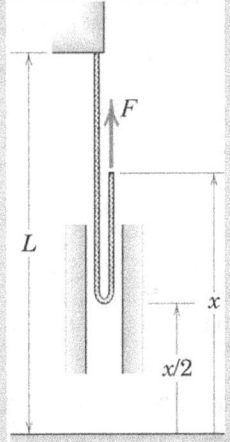

FIGURE P5

6. ●●●Replace the rope of Prob. 5 by an open-link chain with the same mass per unit length. The free end is given a constant upward velocity v. Write expressions for F, the tension T_1 at the bottom of the moving part, and the force R supporting the fixed end in terms of x. Also find the energy loss Q in terms of x.

7. ●●●The flexible non-extensible rope of length $\pi r/2$ and mass ρ per unit length is attached at A to the fixed quarter-circular guide and allowed to fall from rest in the horizontal position. When the rope comes to rest in the dashed position, the system will have lost energy. Determine the loss ΔQ and explain what becomes of the lost energy

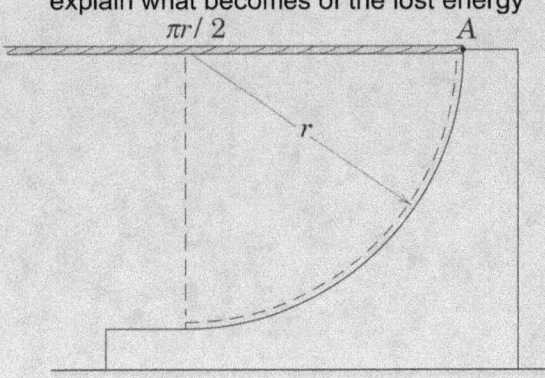

FIGURE P7

8. ●●●The chain of length L and mass ρ per unit length is released from rest on the smooth horizontal surface with a negligibly small overhang x to initiate motion. Determine (a) the acceleration a as a function of x, (b) the tension T in the chain at the smooth corner as a function of x, and (c) the velocity v of the last link A as it reaches the corner

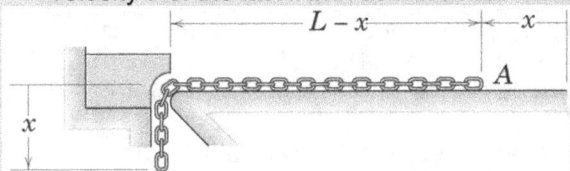

FIGURE P8

9. ●●●The chain of mass ρ per unit length passes over the small freely turning pulley and is released from rest with only a small imbalance h to initiate motion. Determine the acceleration a and velocity v of the chain and the force R supported by the hook at A, all in terms of h as it varies from essentially zero to H. Neglect the weight of the pulley and its supporting frame and the weight of the small amount of chain in contact with the pulley. (*Hint:* The force R does not equal two times the equal tensions T in the chain tangent to the pulley.)

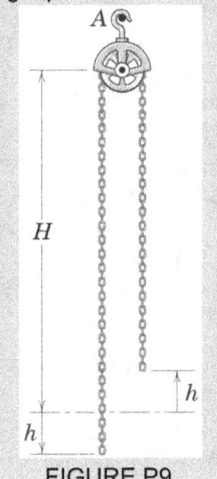

FIGURE P9

14. CENTRE OF MASS (CM)

FIGURE 1 (a) The center of mass of a boomerang is a point outside of the boomerang. (b) The parabolic path followed by the center of mass when a hammer is tossed through the air. (c) British pole-vaulter Ben Challenger's center of mass actually passes *beneath* the bar as his body passes over the bar.

We define the **center of mass** (CM) of a system of particles (such as a person) in order to predict the possible motion of the system.

The center of mass of a system of particles is the point that moves as though (1) all of the system's mass were concentrated there and (2) all external forces were applied there.

The CM of an object is not necessarily located within the object; for some objects, such as a boomerang, the center of mass is located outside of the object itself (Fig. 1a).

What if a system is not isolated, but has external interactions? Again, imagine all of the mass of the system concentrated into a single point particle located at the CM. The motion of this fictitious point particle is determined by Newton's second law, where the net force is the sum of all of the external forces acting on *any part* of the system. In the case of a complex system composed of many parts interacting with each other, the motion of the CM is considerably simpler than the motion of an arbitrary particle of the system (Fig.1b ,c).

Here we discuss how to determine where the center of mass of a system of particles is located. We start with a system of only two particles, and then we consider a system of a many number of particles (a solid body, such as a baseball bat). Later in the chapter, we discuss how the center of mass of a system moves when external forces act on the system.

15. SYSTEMS OF PARTICLES

15.1. TWO PARTICLES

EXAMPLE 31. For a system composed of two particles, the center of mass lies somewhere on a line between the two particles. In Fig. 2, particles of masses m_1 and m_2 are located at positions x_1 and x_2, respectively. We define the location of the cm for these two particles as-

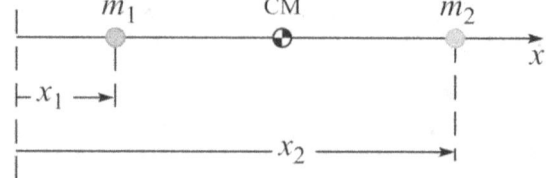

FIGURE 2. Two particles of equal mass located at positions x_1 and x_2 from the origin. The cm is midway between the two.

$$x_{CM} = \frac{m_1 x_1 + m_2 x_2}{m_1 + m_2}$$

If $m_1 + m_2 = M$, is the total mass of the system, then

$$x_{CM} = \frac{m_1 x_1 + m_2 x_2}{M}$$

If the two masses are equal ($m_1 = m_2 = m$), then x_{CM} is midway between them, since in this case

$$x_{CM} = \frac{m(x_1+x_2)}{2m} = \frac{(x_1+x_2)}{2}$$

If one mass is greater than the other, say, $m_1 > m_2$, then the CM is closer to the larger mass (Fig.3). If all the mass is concentrated at x_2, so $m_1 = 0$, then
$x_{CM} = \frac{0x_1+m_2x_2}{0+m_2} = \frac{m_2x_2}{m_2} = x_2$, as we would expect.

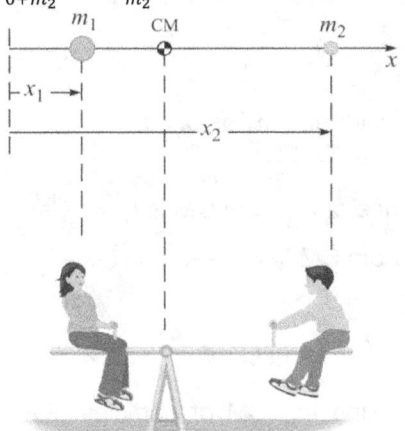

FIGURE.3 Two particles of unequal mass. The cm is closer to the more massive particle. For two children balanced on a see-saw, the cm is at the fulcrum.

15.2. MANY PARTICLES

Now let us consider a system consisting of n particles, where n could be very large. This system could be an extended object which we consider as being made up of n tiny particles. If these n particles are all along a straight line (call it the x axis), we define the CM of the system to be located at-

$$x_{CM} = \frac{m_1x_1+m_2x_2+m_3x_3+\cdots m_nx_n}{m_1+m_2+m_3+\cdots+m_n} = \frac{1}{M}\sum_{i=1}^{n} m_i x_i \quad \ldots (1)$$

where $m_1, m_2, m_3, \ldots, m_n$ are the masses of each particle and $x_1, x_2, x_3, \ldots, x_n$ are their positions. The symbol $\sum_{i=1}^{n}$ is the summation sign meaning to sum over all the particles, where i takes on integer values from 1 to n. (Often, we simply write $\sum m_i x_i$, leaving out the $i = 1$ to n). The total mass of the system is $M = \sum m_i$.

Three Dimensions.

If the particles are distributed in three dimensions, the center of mass must be identified by three coordinates. By extension of Eq. 1, they are

$$\left. \begin{array}{l} x_{CM} = \frac{1}{M}\sum_{i=1}^{n} m_i x_i \\ y_{CM} = \frac{1}{M}\sum_{i=1}^{n} m_i y_i \\ z_{CM} = \frac{1}{M}\sum_{i=1}^{n} m_i z_i \end{array} \right\} \quad \ldots (2)$$

Now, the position vector of the center of mass is-
$$\vec{r}_{CM} = x_{CM}\hat{i} + y_{CM}\hat{j} + z_{CM}\hat{k}$$
or $\quad \vec{r}_{CM} = \frac{m_1\vec{r}_1+m_2\vec{r}_2+m_3\vec{r}_3+\cdots m_n\vec{r}_n}{m_1+m_2+m_3+\cdots m_n}$

or $\quad \vec{r}_{CM} = \frac{1}{M}\sum_{i=1}^{n} m_i \vec{r}_i \quad \ldots (3)$

here $\vec{r}_1, \vec{r}_2, \vec{r}_3, \ldots, \vec{r}_n$ are position vectors of point masses $m_1, m_2, m_3, \ldots, m_n$ respectively.

> C.M. is an imaginary point, which may or may not be located on the system, and in many (emphasize not always) cases of mechanics, whole mass can be assumed to be concentrated on it.

Illustration: If you throw a rotating pen in air, only one point (CM) will be going as parabola.

16. CM OF EXTENDED, CONTINIOUS OBJECTS

Although locating the center of mass for an extended, continuous object is somewhat more cumbersome than locating the center of mass of a small number of particles, the basic ideas we have discussed still apply. Think of an extended object as a system containing a large number of small mass elements such as the cube in FIGURE 4. Because the separation between elements is very small, the object can be considered to have a continuous mass distribution. By dividing the object into elements of mass Δm_i with coordinates (x_i, y_i, z_i) we see that the x coordinate of the center of mass is approximately

$x_{CM} \approx \frac{1}{M}\sum_{i=1}^{n} x_i \Delta m_i$,

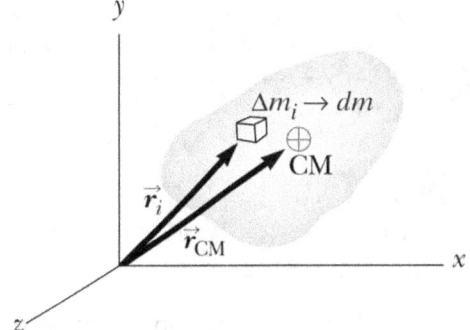

FIGURE 1 An extended object, here shown in only two dimensions, can be considered to be made up of many tiny particles (n), each having a mass Δm_i. One such particle is shown located at a point $\vec{r}_i = x_i\hat{i} + y_i\hat{j} + z_i\hat{k}$. We take the limit of $n \to \infty$ so Δm_i becomes the infinitesimal dm.

With similar expressions for y_{CM} and z_{CM}. If we let the number of elements n approach infinity, the size of each element approaches zero and x_{CM} is given precisely. In this limit, we replace the sum by an integral and Δm_i by the differential element dm:

$x_{CM} = \lim_{\Delta m_i \to 0} \frac{1}{M}\sum_{i=1}^{n} x_i \Delta m_i = \frac{1}{M}\int x\, dm$

Likewise, for y_{CM} and z_{CM} we obtain

$y_{CM} = \frac{1}{M}\int y\, dm$ and $z_{CM} = \frac{1}{M}\int z\, dm$

$$x_{CM} = \frac{1}{M}\int x\, dm,\ y_{CM} = \frac{1}{M}\int y\, dm,\ z = \frac{1}{M}\int z\, dm \quad \ldots(3)$$

or $\quad x_{CM} = \frac{\int x\, dm}{\int dm},\ y_{CM} = \frac{\int y\, dm}{\int dm},\ z = \frac{\int z\, dm}{\int dm}$

where $M = \int dm$ is now the mass of the object.

Evaluating these integrals for most common objects (such as a television set or an elephant) would be difficult, so here we consider only *uniform* objects. Such objects have uniform *density*, or mass per unit volume; that is, the density ρ is the same for any given element of an object as for the whole object.

Density $\quad \rho = \frac{dm}{dV} = \frac{M}{V} \quad \ldots(4)$

where dV is the volume occupied by a mass element dm, and V is the total volume of the object. Substituting $dm = \frac{M}{V}dV$ from Eq. 4 into Eq. 3 gives

$x_{CM} = \frac{1}{V}\int x\, dV,\ y_{CM} = \frac{1}{V}\int y\, dV, z = \frac{1}{V}\int z\, dV \quad \ldots(5)$

We can express the vector position of the center of mass of an extended object in the form

$\vec{r}_{cm} = \frac{\int \vec{r}\, dm}{\int dm} = \frac{1}{M}\int \vec{r}\, dm$

$\vec{r}$ is position vector (PV) of dm with respect to origin and M is total mass of system.

> ☞ A concept similar to center of mass is **center of gravity (CG)**. The CG of an object is that point at which the force of gravity can be considered to act. The force of gravity actually acts on all the different parts or particles of an object, but for purposes of determining the translational motion of an object as a whole, we can assume that the entire weight of the object (which is the sum of the weights of all its parts) acts at the CG. There is a conceptual difference between the center of gravity and the center of mass, but for nearly all practical purposes, they are at the same point.
> ☞ There would be a difference between the CM and CG only in the unusual case of an object so large that the acceleration due to gravity, g, was different at different parts of the object.

Symmetry as a Shortcut. You can bypass one or more of these integrals if an object has a point, a line, or a plane of symmetry. The center of mass of such an object then lies at that point, on that line, or in that plane. For example, the center of mass of a uniform sphere (which has a point of symmetry) is at the center of the sphere (which is the point of symmetry). The center of mass of a uniform cone (whose axis is a line of symmetry) lies on the axis of the cone. The center of mass of a banana (which has a plane of symmetry that splits it into two equal parts) lies somewhere in the plane of symmetry.

EXAMPLE 32. Distance between m_1 and m_2 is l. Find distance between m_1 and C.M.

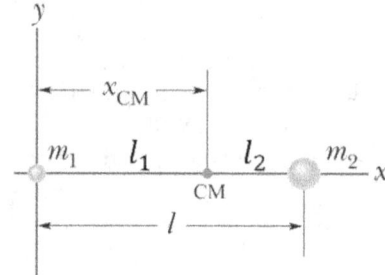

SOLUTION Considering m_1 as origin,

$x_{CM} = \frac{m_1 x_1 + m_2 x_2}{m_1 + m_2} = \frac{m_1 0 + m_2 l}{m_1 + m_2} = \frac{m_2 l}{m_1 + m_2}$

Thus, the distance of m_1 from CM is $l_1 = \frac{m_2 l}{m_1 + m_2}$

distance of m_2 from CM $l_2 = l - x_{cm} = l - \frac{m_2 l}{m_1 + m_2} = \frac{m_1 l}{m_1 + m_2}$

$\Rightarrow \frac{l_1}{l_2} = \frac{m_2}{m_1}$

☞ C.M. divides two point masses in inverse ratio of their masses

EXAMPLE 33. Find the CM of particles placed at the vertices of an equilateral triangle of side a.

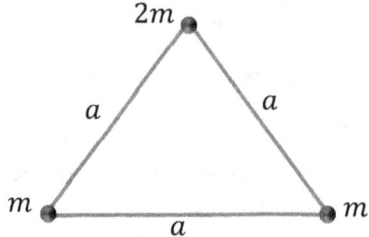

SOLUTION Following diagram shows the system of particles with their respective coordinates-

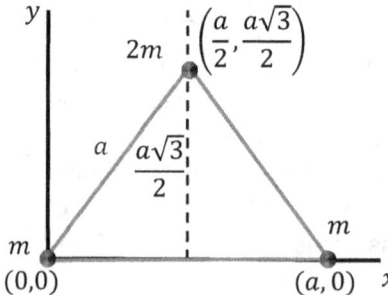

Applying the definition of CM for each coordinate, we have-

$x_{CM} = \frac{m_1 x_1 + m_2 x_2 + m_3 x_3}{m_1 + m_2 + m_3} = \frac{m \times 0 + ma + 2m \times \frac{a}{2}}{4m} = \frac{a}{2}$

$y_{CM} = \frac{m_1 y_1 + m_2 y_2 + m_3 y_3}{m_1 + m_2 + m_3} = \frac{m \times 0 + 2m \times a\frac{\sqrt{3}}{2}}{4m} = \frac{a\sqrt{3}}{4}$

Therefore, coordinates of CM $\left(\frac{a}{2}, \frac{a\sqrt{3}}{4}\right)$

EXAMPLE 34. Regular hexagon Six masses are placed at the vertices of a regular hexagon as shown in adjoining figure. Find the position of CM.

LINEAR MOMENTUM, IMPULSE AND COLLISIONS

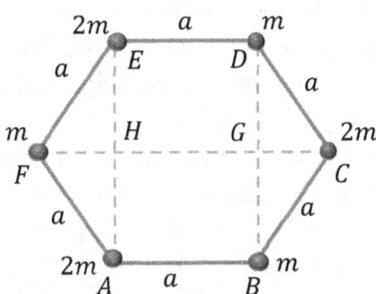

APPROACH Masses at A & E can be placed at centre of AE, similarly masses at B & D can be placed at centre of BD. Now calculate the position of CM of all the masses placed on the line FC.

SOLUTION Symbolic solution is given in following diagrams. In each diagram, we have replaced two masses by their resultant mass at their CM.

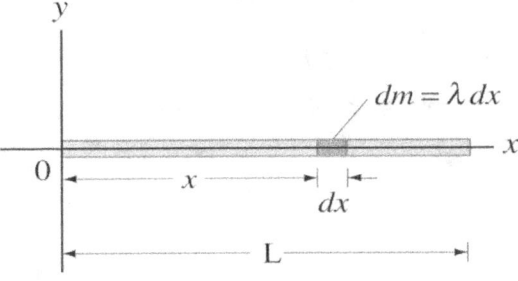

Therefore, location of CM from A is given by, $x_{CM} = \frac{a}{2}$, $y_{CM} = AH = \frac{a}{2}\sqrt{3}$

☞ While calculating C.M. we can replace bodies by their C.M.

EXAMPLE 35. CM OF A THIN ROD (a) Show that the CM of a uniform thin rod of length L and mass M is at its center. (b) Determine the CM of the rod assuming its linear mass density (its mass per unit length) varies linearly from $\lambda = \lambda_0$ at the left end to double that value, $\lambda = 2\lambda_0$, at the right end.

APPROACH We choose a coordinate system so that the rod lies on the x axis with the left end at $x = 0$, Fig. 1. Then $y_{CM} = 0$ and $z_{CM} = 0$.

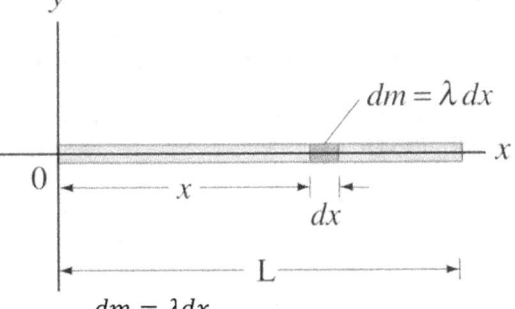

SOLUTION (a) The rod is uniform, so its mass per unit length (linear mass density λ) is constant and we write it as $\lambda = \frac{M}{L}$. We now imagine the rod as divided into infinitesimal elements of length dx, each of which has mass

$$dm = \lambda \, dx = \frac{M}{L} dx.$$

$$\therefore \quad x_{cm} = \frac{1}{M} \int x \, dm = \frac{1}{M} \int_0^L \frac{M}{L} x \, dx = \frac{L}{2}.$$

This result x_{CM} at the center, is what we expected.

(b) Now we have $\lambda = \lambda_0$ at $x = 0$ and we are told that λ increases linearly to $\lambda = 2\lambda_0$ at $x = L$, so we write

$$\lambda = \lambda_0(1 + \alpha x)$$

which satisfies $\lambda = \lambda_0$ at $x = 0$, increases linearly, and gives $\lambda = 2\lambda_0$ at $x = L$ if $(1 + \alpha x) = 2$. In otherwords, $\alpha = \frac{1}{L}$.

$$\therefore \quad \lambda = \lambda_0\left(1 + \frac{1}{L}x\right)$$

Now, $x_{CM} = \frac{1}{M}\int x\, dm = \frac{1}{M}\int_0^L x\lambda_0\left(1 + \frac{1}{L}x\right) dx = \frac{\lambda_0}{M}\left[\frac{x^2}{2} + \frac{x^3}{3L}\right]_0^L = \frac{5\lambda_0}{6M}L^2.$

We can write M in terms of λ_0 and L, as

$$M = \int_{x=0}^L dm = \int_{x=0}^L \lambda \, dx = \lambda_0 \int_0^L \left(1 + \frac{1}{L}x\right) dx$$

$$= \lambda_0\left[x + \frac{1}{2L}x^2\right]_0^L = \frac{3}{2}\lambda_0 L$$

Then, $x_{CM} = \frac{5\lambda_0}{6M}L^2 = \frac{5}{9}L.$

which is more than halfway along the rod, as we would expect since there is more mass to the right.

EXAMPLE 36. Determine the CM of the rod assuming its linear mass density (its mass per unit length) varies as $\lambda = kx^2$

SOLUTION Let us consider an elementary part of length dx on the rod. If dm is the mass of this small length, then

$$dm = \lambda dx$$

$\because \lambda = kx^2 \therefore dm = \lambda dx = kx^2 dx$

Total mass of the rod,

$$M = k\int_0^L x^2 dx = \frac{kL^3}{3}$$

$$\therefore \quad x_{CM} = \frac{\int x\, dm}{M} = \frac{k\int_0^L x^3 dx}{\frac{kL^3}{3}} = \frac{3L}{4}$$

EXAMPLE 37. Determine the CM of the uniform thin L-shaped construction brace shown in Fig. 1.

30 CONCEPTS AND PROBLEMS IN PHYSICS

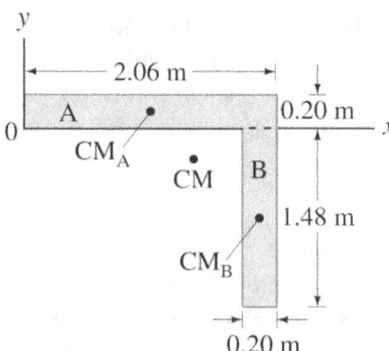

APPROACH We can consider the object as two rectangles: rectangle A, which is $2.06\ m \times 0.20\ m$, and rectangle A, which is $1.48\ m \times 0.20\ m$, and rectangle B, which is $1.48\ m \times 0.20\ m$. We choose the origin at 0 as shown. We assume a uniform thickness t.

SOLUTION CM of rectangle A is at $x_A = 1.03\ m$, $y_A = 0.10\ m$.

The CM of B is at $x_B = 1.96\ m$, $y_B = -0.74\ m$.

The mass of A, whose thickness is t, is

$M_A = (2.06\ m)(0.20\ m)(t)(\rho) = (0.412\ m^2)(\rho t),$

where ρ is the density (mass per unit volume). The mass of B is

$M_B = (1.48\ m)(0.20\ m)(t)(\rho) = (0.296\ m^2)(\rho t),$

and the total mass is $M = M_A + M_B = 0.708\ m^2 (\rho t)$. Thus

$$x_{CM} = \frac{M_A x_A + M_B x_B}{M}$$

$$= \frac{(0.412\ m^2)(1.03\ m)+(0.296\ m^2)(1.96\ m)}{0.708\ m^2} = 1.42\ m$$

where ρt was canceled out in numerator and denominator. Similarly,

$$y_{CM} = \frac{M_A y_A + M_B y_B}{M}$$

$$= \frac{(0.412\ m^2)(0.10\ m)+(0.296\ m^2)(-0.74\ m)}{0.708\ m^2} = -0.25\ m,$$

which puts the CM approximately at the point so labeled in Fig. 1. In thickness, $z_{CM} = t/2$, since the object is assumed to be uniform.

This is an example in which the CM can actually lie *outside* the object.

EXAMPLE 38. Find the CM of L shaped rod system shown in adjoining figure-

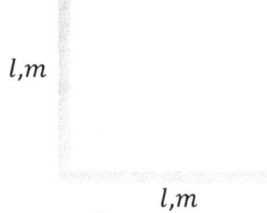

SOLUTION

$x_{cm} = \frac{ml/2+0}{2m} = \frac{l}{4}$

$y_{cm} = \frac{0+m\frac{l}{2}}{2m} = \frac{l}{4}$

EXAMPLE 39. Consider regular polygons with number of sides $n = 3, 4, 5,$ ... as shown in the figure. The center of mass of all the polygons is at height h from the ground. They roll on a horizontal surface about the leading vertex without slipping and sliding as depicted. The maximum increase in height of the locus of the center of mass for each polygon is Δ. Then Δ depends on n and h as **[IIT 2017]**

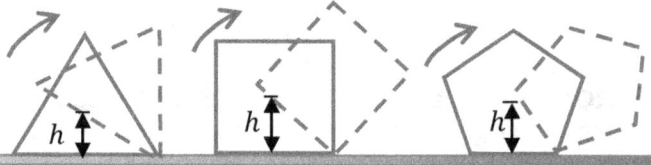

(A) $\Delta = h \sin^2(\pi/n)$ (B) $\Delta = h \sin(2\pi/n)$
(C) $\Delta = h \tan^2(\pi/2n)$ (D) $\Delta = h\left(\frac{1}{\cos(\pi/n)} - 1\right)$

SOLUTION Let n be the number of sides of a regular polygon. By symmetry, its centre of mass O will be equidistant from each vertex i.e., it lies at the centre of the circumscribed circle. Let r be the radius of circumscribed circle and h be the perpendicular distance of O from any side (see figure). The angle subtended by any side on the centre O is $2\pi/n$ and $\angle POQ = \pi/n$.

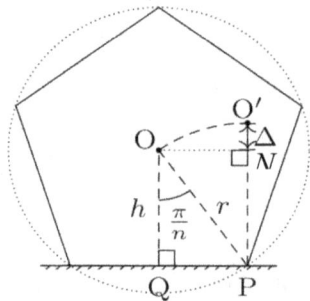

When polygon rolls about the vertex P (without slipping or sliding), the point O moves in a circle of radius r centered at P. The point O reaches the maximum height (point O' in the figure) when PO' is perpendicular to PQ. Thus, the maximum increase in height of the locus of the centre of mass O is given by

$$\Delta = r - h = \frac{h}{\cos(\pi/n)} - h = h\left(\frac{1}{\cos(\pi/n)} - 1\right)$$

Therefore, option D is correct.

17. OPTIONAL DERIVATIONS

17.1. CM OF A HALF RING

EXAMPLE 40. A thin strip of material is bent into the shape of a semicircle of radius R (Fig. 1a). Find its center of mass.

LINEAR MOMENTUM, IMPULSE AND COLLISIONS 31

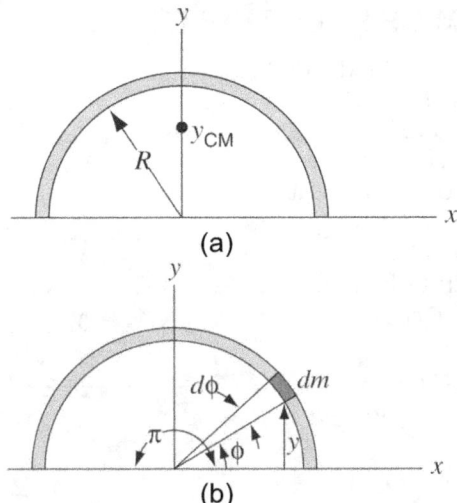

FIGURE 1. (a) A thin strip of metal bent into the shape of a semicircle. (b) An element of the strip of mass dm located at the angular coordinate ϕ.

APPROACH This case has symmetry about the y axis (that is, for every particle to the left of the y axis there is a particle in a similar location to the right of the y axis). The center of mass must therefore lie on the y axis; that is, However, $x_{CM} = 0$, However, there is no symmetry about the x axis, so we must use equation $y_{CM} = \frac{\int y \, dm}{\int dm}$ to find y_{CM}. Using an angular coordinate simplifies the integration to be performed.

SOLUTION Consider the small element of mass dm shown in Fig. 1b. It subtends an angle $d\phi$, and since the total mass M of the strip subtends an angle π (a full circle would subtend an angle 2π), the mass dm must be the same fraction of M as $d\phi$ is of π. That is, $\frac{dm}{M} = \frac{d\phi}{\pi}$ or $dm = \frac{M}{\pi} d\phi$. The element dm is located at the coordinate $y = R \sin \phi$.

$$\therefore \quad y_{CM} = \frac{\int y \, dm}{\int dm}$$
$$= \frac{\frac{M}{\pi} \int_0^\pi R \sin \phi \, d\phi}{\frac{M}{\pi} \int_0^\pi d\phi}$$
$$= \frac{R[-\cos \phi]_0^\pi}{\pi} = -\frac{R}{\pi}[\cos \pi - \cos 0] = \frac{2R}{\pi} = 0.637R.$$

i.e., $y_{CM} = \frac{2R}{\pi}$, $x_{CM} = 0$

The center of mass is roughly two-thirds of a radius along the y axis. Note that, as this case illustrates, the center of mass does not need to be within the volume or the material of an object.

17.2. CM OF A HALF DISC

EXAMPLE 41. Find the CM of half disk of radius R and mass m. The mass density of the half disk is uniform.

APPROACH To find the position of CM, we divide the whole half disk into very thin concentric semicircular rings. One such elementary thin semicircular ring of radius r and thickness dr is shown in the figure. For this ring, its whole mass can be considered at its CM, which is at $y = -\frac{2r}{\pi}$ on the Y axis.

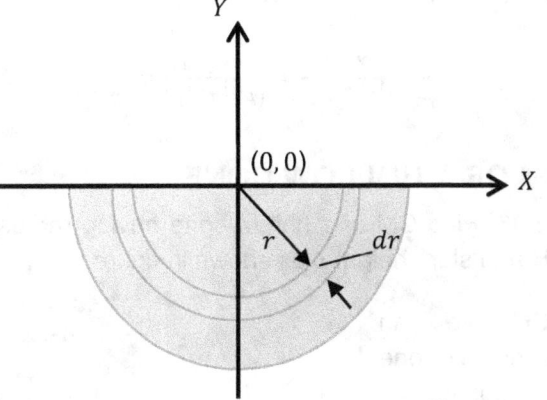

Now, mass density of the disk $\sigma = \frac{m}{\pi R^2/2}$

∴ mass of the elementary ring, $dm = \left(\frac{m}{\pi R^2/2}\right) \times 2\pi r \, dr = -\frac{4m}{R^2} dr$

Now by using formula $y_{CM} = \frac{\int y \, dm}{\int dm}$ with limits $r = 0$ to $r = R$, we can find y_{CM}. BY symmetry it is clear that $x_{CM} = 0$.

SOLUTION $y_{CM} = \frac{\int y \, dm}{\int dm} = \frac{\int_0^R \left(-\frac{2r}{\pi}\right)\frac{4m}{R^2} r \, dr}{\int_0^R \frac{4m}{R^2} r \, dr}$

$$= -\frac{\frac{8m}{\pi R^2} \int_0^R r^2 \, dr}{\frac{4m}{R^2} \int_0^R r \, dr} = -\frac{2}{\pi} \frac{\left[\frac{R^3}{3}\right]}{\frac{R^2}{2}} = -\frac{4R}{3\pi}$$

17.3. CM OF A SOLID CONE

EXAMPLE 42. Find CM of a solid cone having radius R and height H as shown in figure.

APPROACH We can consider a solid cone as a continuous collection of elementary discs having radius range from 0 to R. The mass of such an elemental disc at a vertical distance y from the vertex of the cone is given by-
$dm = (\pi r^2 dy)\rho$
Here, ρ is the mass density of the material of the cone.
Here, we have two variable r and y, so we have to convert r in terms of variable y.
From adjoining figure,
$\frac{r}{y} = \frac{R}{H} \Rightarrow r = \frac{R}{H} y$

$\therefore dm = \pi \left(\frac{R}{H} y\right)^2 dy \rho$

Now, net mass of cone $M = \int_0^H \pi \left(\frac{R}{H} y\right)^2 dy \rho$

$$= \rho \frac{\pi R^2}{H^2} \left[\frac{y^3}{3}\right]_0^H = \rho \frac{\pi R^2}{H^2} \frac{H^3}{3} = \rho \frac{1}{3}\pi R^2 H$$

From symmetry, we can write $x_{CM} = 0$

To calculate y_{CM}, we use $y_{CM} = \frac{\int_0^H y(dm)}{M}$

SOLUTION $y_{CM} = \frac{\int_0^H y\pi\left(\frac{R}{H}y\right)^2 dy\rho}{\left(\rho\times\frac{1}{3}\pi R^2\right)H} = \frac{\rho\frac{\pi R^2}{H^2}\left[\frac{y^4}{4}\right]_0^H}{\left(\rho\times\frac{1}{3}\pi R^2\right)H} = \frac{3H}{4}$

$x_{CM} = 0$

17.4. CM OF A HOLLOW CONE

EXAMPLE 43. Find CM of a hollow cone having radius R, height H and slant height L as shown in figure.

APPROACH We can consider a hollow cone as a continuous collection of elementary rings having radius range from 0 to R. The mass of such an elemental ring at slant distance l shown in adjoining figure is given by-

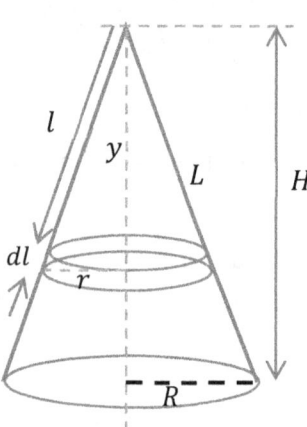

$dm = \sigma \times (2\pi r\, dl)$

Here, σ is the surface mass density of the hollow cone. Here, we have two variables l and r, so we first we have to convert r in terms of variable l.
From adjoining figure, we have
$\frac{r}{l} = \frac{R}{L} \Rightarrow r = \frac{R}{L}l$

$\therefore \quad dm = \sigma \times (2\pi r\, dl) = 2\pi\sigma\left(\frac{R}{L}l\right)dl$

Now, net mass of cone $M = \int_0^L 2\pi\sigma\left(\frac{R}{L}l\right)dl$

$= \frac{2\pi\sigma R}{L}\left[\frac{l^2}{2}\right]_0^L = \sigma\pi RL$

From symmetry of the figure, $x_{CM} = 0$

And y_{CM} is given by $y_{CM} = \frac{\int_0^H y(dm)}{M}$

Now, $ydm = \sigma \times y(2\pi r\, dl)$

Here we have to convert y and r in terms of l.
From the figure, we have, $\frac{y}{l} = \frac{H}{L}$ or $y = \frac{H}{L}l$
and $r = \frac{R}{L}l$

$\therefore \quad ydm = \sigma \times y(2\pi r\, dl) = 2\pi\sigma\left(\frac{H}{L}l\right)\left(\frac{R}{L}l\right)dl$

Now, $\int_0^H y(dm) = \int_0^L 2\pi\sigma\left(\frac{H}{L}l\right)\left(\frac{R}{L}l\right)dl$

$= \frac{2\pi\sigma RH}{L^2}[l^3/3]_0^L = \frac{2}{3}\sigma\pi RHL$

SOLUTION $y_{CM} = \frac{\int_0^H y(dm)}{M} = \frac{\frac{2}{3}\sigma\pi RHL}{\sigma\pi RL} = \frac{2}{3}H$, $x_{CM} = 0$

17.5. CM OF HALF SHELL

EXAMPLE 44. Find the CM of half shell as shown in figure

APPROACH Considering an element at angle θ and of thickness $R\, d\theta$.
Radius of this element
$r = R\cos\theta$
Mass of element
$dm = \sigma \cdot 2\pi rR\, d\theta$
$r = R\cos\theta$

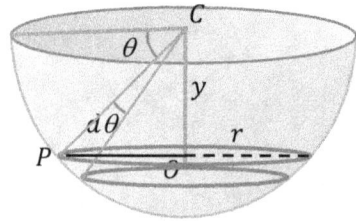

SOLUTION
$y_{CM} = \frac{\int y\, dm}{\int dm} =$

$\frac{\int_0^{\pi/2} R\sin\theta \sigma 2\pi rR\, d\theta}{\int_0^{\pi/2} \sigma \cdot 2\pi rR\, d\theta}$

$= \frac{\sigma 2\pi R^2 \int R\sin\theta \sigma 2\pi rR\, d\theta}{-2\pi R\int R\cos\theta\, d\theta}$

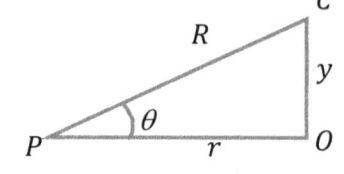

$y_{cm} = \frac{R\int_0^{\pi/2}\sin\theta\cos\theta\, d\theta}{\int_0^{\pi/2}\cos\theta\, d\theta} = \frac{R}{2}\frac{\int_0^{\pi/2}\sin 2\theta\, d\theta}{\int_0^{\pi/2}\cos\theta\, d\theta}$

$= \frac{R}{2}\frac{\left[-\frac{\cos 2\theta}{2}\right]_0^{\pi/2}}{[\sin\theta]_0^{\pi/2}} = \frac{R}{4}\frac{[1-(-1)]}{[1-0]} = \frac{R}{2}$

18. THE CENTRE OF MASS AFTER REMOVAL OF A PART OF A BODY

If a portion of a body is taken out, the remaining portion may be considered as, original mass (M) - mass of the removed part (m).

= {Original mass (M)}
  + {−mass of the removed part (m)}

The Formula changes to;
$x_{cm} = \frac{Mx - mx'}{M-m}$, $y_{cm} = \frac{My - my'}{M-m}$,
Where primed ones represent the coordinate of the CM of the removed part.

EXAMPLE 45. A thin homogeneous lamina is in the form of a circular disc of radius R. From it a circular hole is cut off exactly half the radius of the lamina and touching the lamina's circumference. Find the centre of mass of the remaining part.

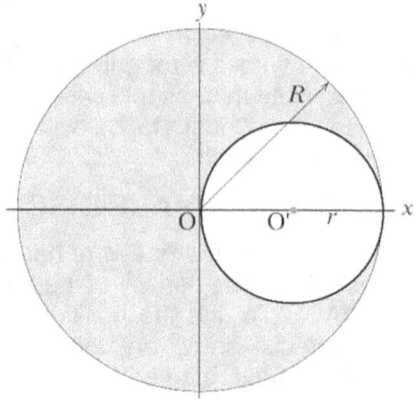

SOLUTION Let the centre of the

lamina be the origin. Due to symmetry, the CM will be on the x-axis. Let m be the mass of the circular lamina. Then mass m' of the removing part is given by

$$m' = \frac{M}{\pi R^2}(\pi r^2) = \frac{M}{R^2}\left(\frac{R}{2}\right)^2 = \frac{M}{4}$$

$$x_{cm} = \frac{M.0 - m'.r}{M - m'}$$

The negative sign of m' denotes that it has been removed.

$$x_{cm} = \frac{\frac{MR}{4\,2}}{M - \frac{M}{4}} = -\frac{R}{6} \qquad \left[\because r = \frac{R}{2}\right]$$

Thus, the centre of mass of the remaining part lies at a distant $\frac{R}{6}$, towards left of the origin. (i.e. the initial centre of mass of the disc)

19. MOTION OF THE CENTER OF MASS

Now that we know how to find the position of the CM of a system, we turn our attention to the motion of the CM. How is the velocity of the CM related to the velocities of the various parts of the system?

During a short time interval Δt, the displacement of the ith particle is $\Delta \vec{r}_i = \vec{v}_i \Delta t$ and the displacement of the center of mass is $\Delta \vec{r}_{CM} = \vec{v}_{CM} \Delta t$. From the definition of the CM, the displacements must be related as follows:

$$\Delta \vec{r}_{CM} = \frac{\sum_i m_i \Delta \vec{r}_i}{M} \quad \ldots (1)$$

(Displacement in CM of a system of particles)

or $\quad \Delta \vec{r}_{CM} = \frac{m_1 \Delta \vec{r}_1 + m_1 \Delta \vec{r}_2 + \ldots + m_n \Delta \vec{r}_n}{m_1 + m_2 + \ldots + m_n}$

In terms of x, y and z coordinates-

$\Delta x_{CM} = \frac{m_1 \Delta x_1 + m_2 \Delta x_2 + \ldots}{m_1 + m_2 + \ldots}$, $\Delta y_{CM} = \frac{m_1 \Delta y_1 + m_1 \Delta y_1 + \ldots}{m_1 + m_2 + \ldots}$

and $\Delta z_{CM} = \frac{m_1 \Delta z_1 + m_1 \Delta z_1 + \ldots}{m_1 + m_2 + \ldots}$

where Δx_{CM}, Δy_{CM} and Δz_{CM} is the change in the co-ordinates of the CM of the system and Δx_1, Δx_2, ...; Δy_1, Δy_2, ... and Δz_1, Δz_2, ... are change in the positions of the of the individual masses.

From equation (1),

$\Delta \vec{r}_{CM} = \frac{\sum_i m_i \Delta \vec{r}_i}{M} \Rightarrow \vec{v}_{CM} \Delta t = \frac{\sum_i m_i \vec{v}_i \Delta t}{M}$

or $\quad \vec{v}_{CM} = \frac{\sum_i m_i \vec{v}_i}{M}$

$\left[\text{or} \quad \vec{v}_{CM} = \frac{m_1 \vec{v}_1 + m_2 \vec{v}_2 + m_3 \vec{v}_3 + \cdots + m_n \vec{v}_n}{m_1 + m_2 + m_3 + \cdots + m_n}\right]$

or $\quad \vec{v}_{CM} = \frac{\sum_i m_i \vec{v}_i}{M}$

$\left[\text{or} \quad \vec{v}_{CM} = \frac{m_1 \vec{v}_1 + m_2 \vec{v}_2 + m_3 \vec{v}_3 + \cdots + m_n \vec{v}_n}{m_1 + m_2 + m_3 + \cdots + m_n}\right]$

Multiplying both sides by M, we get

$M \vec{v}_{CM} = \sum_i m_i \vec{v}_i \quad \ldots (2)$

The right side of Eq. (2) is the sum of the momenta of the particles that constitute the system—the total momentum of the system $\vec{P}$. i.e.,

$\vec{P} = m_1 \vec{v}_1 + m_2 \vec{v}_2 + m_3 \vec{v}_3 + \cdots + m_n \vec{v}_n$
$= \vec{p}_1 + \vec{p}_2 + \vec{p}_3 + \cdots + \vec{p}_n$

Therefore,

$$\vec{P} = M \vec{v}_{CM} \quad \ldots (3)$$

(linear momentum of system of particles)

For two-dimensional motion, Eq. (3) is equivalent to two component equations

$P_x = M v_{CM,x} \quad$ and $\quad P_y = M v_{CM,y}$

Now we know that, for an isolated system, the total linear momentum is conserved. In such a system, Eq. (3) implies that the CM must move with constant velocity regardless of the motions of the individual particles. On the other hand, what if the system is not isolated? If a net external force acts on a system, the CM does not move with constant velocity. Instead, it moves as if all the mass were concentrated there into a fictitious point particle with all the external forces acting on that point.

20. FORCE AND MOMENTUM

If we take the time derivative of Eq. 3 (the velocity can change but not the mass), we find

$\frac{d\vec{P}}{dt} = M \frac{d\vec{v}_{CM}}{dt} = M \vec{a}_{CM}$

where M is the total mass of the system, $\frac{d\vec{v}_{CM}}{dt} = \vec{a}_{CM}$ is the acceleration of CM.

$\because \frac{d\vec{P}}{dt} = \sum \vec{F} = \sum \vec{F}_{ext} + \sum \vec{F}_{int}$

$\therefore \quad \sum \vec{F}_{ext} + \sum \vec{F}_{int} = M \vec{a}_{CM}$

Because of Newton's third law, the internal forces all cancel in pairs, and $\sum \vec{F}_{int} = 0$. What survives on the left side is the sum of only the *external* forces:

$\sum \vec{F}_{ext} = M \vec{a}_{CM}$ = net external force acting on the system of particles. Thus

$$\sum \vec{F}_{ext} = \frac{d\vec{P}}{dt} = M \vec{a}_{CM} \quad \ldots (4)$$

(body or collection of particles)

(Newton's second law for a system of particles)

That is, the system translates as if all its mass were concentrated at the CM and all the external forces acted at that point. We can thus treat the translational motion of any object or system of objects as the motion of a particle.

$\vec{R} = \frac{1}{M} \sum m_i \vec{r}_i$ is correct in inertial as well as in non-inertial frames whereas, $\vec{a}_{CM} = \vec{F}/M$ is correct only in inertial frame.

This result is central to the whole subject of mechanics. In fact, we've been using this result all along; without it, we would not be able to represent an extended body as a point particle when we apply Newton's laws. It explains why only *external* forces can affect the motion of an extended body. If you pull upward on your belt, your belt exerts an equal downward force on your hands; these are *internal* forces that cancel and have no effect on the overall motion of your body.

Suppose a cannon shell travelling in a parabolic trajectory (neglecting air resistance) explodes in flight, splitting into two fragments with equal mass (Fig. 1). The fragments follow new parabolic paths, but the centre of mass continues on the original parabolic trajectory, just as though all the mass were still concentrated at that point.

Finally, we note that if the net external force $\sum \vec{F}_{ext}$ is zero, Eq. (4) shows that the acceleration $\vec{a}_{CM}$ of the centre of mass is zero. So, the centre-of-mass velocity is constant.

$$\frac{m_1 \vec{v}_1 + m_2 \vec{v}_2}{m_1 + m_2} = 0$$

or $\quad m_1 \vec{v}_1 + m_2 \vec{v}_2 = 0$

or $\quad \dfrac{m_1 \frac{\Delta \vec{r}_1}{\Delta t} + m_2 \frac{\Delta \vec{r}_2}{\Delta t}}{m_1 + m_2} = 0$

or $\quad \dfrac{m_1 \Delta \vec{r}_1 + m_2 \Delta \vec{r}_2}{m_1 + m_2} = 0 \quad$ or $\quad \Delta \vec{r}_{CM} = 0 \quad$ i.e.

shift in the CM will be zero

or $\quad m_1 \Delta \vec{r}_1 + m_2 \Delta \vec{r}_2 = 0$

In terms of x coordinates $\Delta x_{CM} = \dfrac{m_1 \Delta x_1 + m_2 \Delta x_2}{m_1 + m_2} = 0$

or $\quad m_1 \Delta x_1 + m_2 \Delta x_2 = 0$

Similar expressions can also be written for y and z cordinates.

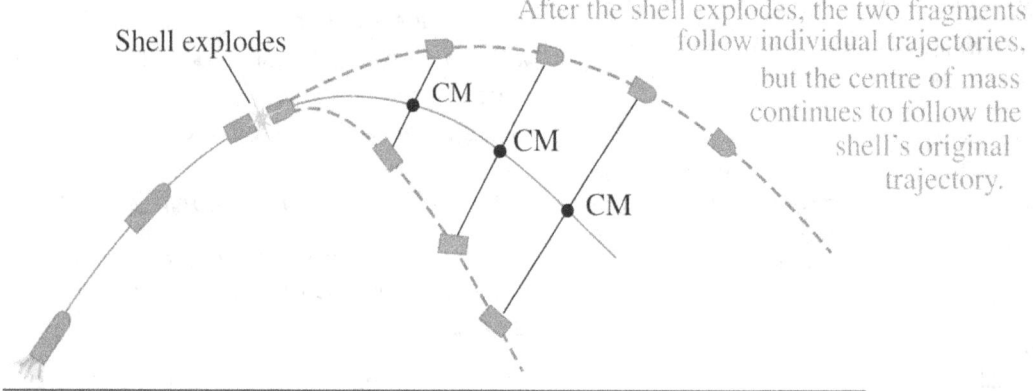

FIGURE 1. A shell explodes into two fragments in flight. If air resistance is ignored, the centre of mass continues on the same trajectory as the shell's path before exploding.

From Eq. (4) the total momentum $\vec{P}$ is also constant. This reaffirms our statement of the principle of conservation of momentum.

21. SHIFT IN CM OF TWO PARTICLE SYSTEM IN ABSENCE OF EXTERNAL FORCE

In the absence of any external force
$\vec{P}$ = constant
i.e., $m\vec{v}_{CM}$ = constant
or $\vec{v}_{CM}$ = constant, since m = constant within the classical limit of mechanics.
Now, *if initially*, $\vec{v}_{CM} = 0$ (note the word 'if'), then in absence of any external force
$\vec{v}_{CM} = \dfrac{\sum_i m_i \vec{v}_i}{M} = 0$
If the problems are asked about two body system in which one is moving over the other in absence of any external force, then we treat the bodies as particle having their whole mass at their respective CM.
Now, for such two particle system, we have

EXAMPLE 46. Two blocks of masses m_1 and m_2 connected by a light spring of stiffness k rest on a smooth horizontal plane. such that the block m_1 touches a vertical wall as shown in the figure. The block m_2 is shifted a small distance x towards the wall and then released. Find the velocity of the centre of mass of the system after block m_1 breaks off the wall.

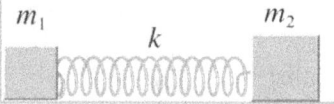

SOLUTION External force acting on the system is the normal reaction of the wall until the spring attains it natural length. Since work done by the normal reaction is zero hence total energy of the system will be conserved. At the moment when spring acquires its natural length velocity of the block m_1 is zero and velocity of the block m_2 is v_2 (say).
Hence from conservation of energy, we get

$$\frac{1}{2}kx^2 = 0 + \frac{1}{2}m_2 v_2^2 \quad \Rightarrow \quad v_2 = \sqrt{\frac{k}{m_2}}x$$

$$v_{CM} = \frac{m_1 v_1 + m_2 v_2}{m_1 + m_2} = \frac{m_1 \times 0 + m_2 \sqrt{k/m_2} x}{m_1 + m_2} = \frac{\sqrt{k m_2} x}{m_1 + m_2}$$

EXAMPLE 47. A dog of mass 10 kg is standing on a flat boat so that it is 20 meters from the shore. It walks 8 m on the boat towards the shore and then stops. The mass of the boat is 40 kg and friction between the boat and the water surface is negligible. How far is the dog from the shore now?

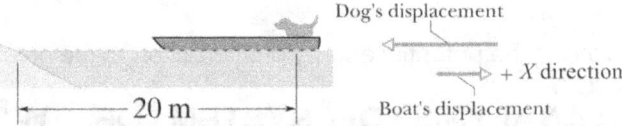

SOLUTION Take boat and dog as a system. Initially centre of mass of the system is at rest. Since no external force is acting on the system, hence centre of mass of the system will remain stationary.

Let Initially distance of the centre of mass of the boat from the shore is x m, then

$$x_{1CM} = \frac{40 \times x + 10 \times 20}{40 + 10} m \qquad \ldots (1)$$

where x_{1CM} = distance of the CM of the system from the shore. Since dog moves towards the shore and centre of mass of the system to be at rest, therefore boat has to move away from the shore. Let distance moved by the boat be x', then

$$x_{2CM} = \frac{40(x + x') + 10(20 - 8 + x')}{40 + 10}$$

$\because x_{1CM} = x_{2CM}$

$\Rightarrow \frac{40x + 200}{50} = \frac{40(x + x') + 10(12 + x')}{50}$

$\Rightarrow 50x' = 80 \quad \Rightarrow \quad x' = 1.6\, m$

Hence distance of the dog from the shore is $(20 - 8 + 1.6)m = 13.6\, m$

For example, in previous case of rotating fan, we can now very easily conclude that momentum is zero, since C.M. will be at axis and have zero velocity.

22. CM SHIFT METHOD (DIRECT METHOD)

Since, there is no external force in horizontal direction, therefore CM will remain at initial state, i.e., in the state of rest.

Now, along X axis, the shift in CM is given by

$$\Delta x_{CM} = \frac{m_1 \Delta x_1 + m_2 \Delta x_2}{m_1 + m_2} = 0$$

$\Rightarrow m_1 \Delta x_1 + m_2 \Delta x_2 = 0 \qquad \ldots (1)$

Let the shift in the position of boat is x away from the shore (i.e., along positive direction of X axis) and

$m_1 = 10\, kg$ (mass of dog), $\Delta x_1 = (-8 + x)m$, $m_2 = 40\, kg$, $\Delta x_2 = +x$, then from (1), we get

$\Rightarrow (10 kg)[(-8 + x)m] + (40 kg)(x\, m) = 0$

$\Rightarrow x = \frac{(10 kg)(8m)}{(50 kg)} = 1.6\, m$

$\Rightarrow$ Boat moves $1.6\, m$ towards right.

Therefore, new distance of dog from the shore $= 20m - 8m + 1.6m = 13.6\, m$

23. C-FRAME

We may attach a frame of reference, designated $X_C Y_C Z_C$, to the center of mass of a system. Relative to this frame, the center of mass is at rest ($v_{CM} = 0$). This is called the **center of mass** or C-frame of reference. In view of equation $P = M v_{CM}$, the total momentum of a system of particles referred to the C-frame of reference is always zero.

$\vec{P} = \sum_i \vec{p}_i = 0$ in the C-frame of reference.

For that reason the C-frame is sometimes also called the **zero momentum frame**. The C-frame is important because many problems can be more simply analyzed in the C-frame compared to ground frame. It is clear that the C-frame moves with a velocity v_{CM} relative to the ground frame. When no external forces act on a system, the C-frame can be considered as inertial.

EXAMPLE 48. The velocities of two particles of masses m_1 and m_2 relative to an inertial observer are v_1 and v_2. Determine the velocity of the center of mass relative to the observer and the velocity of each particle relative to the centre of mass.

SOLUTION From equation $v_{CM} = \frac{\sum_i m_i v_i}{M}$ the velocity of the centre of mass relative to the observer is

$$v_{CM} = \frac{m_1 v_1 + m_2 v_2}{m_1 + m_2}$$

The velocities of each particle relative to the centre of mass, using the Galilean transformation of velocities is

$v_1' = v_1 - v_{CM} = v_1 - \frac{m_1 v_1 + m_2 v_2}{m_1 + m_2} = \frac{m_2(v_1 - v_2)}{m_1 + m_2} = \frac{m_2 v_{12}}{m_1 + m_2}$

and $v_2' = v_2 - v_{CM} = \frac{m_1(v_2 - v_1)}{m_1 + m_2} = -\frac{m_1 v_{12}}{m_1 + m_2}$

where $v_{12} = v_1 - v_2$ is the relative velocity of the two particles. Thus, in the C-frame, the two particles appear to be moving in opposite directions, with velocities inversely proportional to their masses.

24. KINETIC ENERGY OF A SYSTEM

If m_i and v_i are the mass and velocity of i th particle, then KE of system of particle is given by

$KE_{sys} = \sum \frac{1}{2} m_i v_i^2$

Velocity of ith particle with respect to CM
$$v_{i/c} = v_i - v_c$$
or $\quad v_i = v_{i/c} + v_c$

$\therefore KE_{sys} = \sum \frac{1}{2} m_i v_i^2 = \sum \frac{1}{2} m_i (\vec{v}_{i/c} + \vec{v}_c)^2$

$= \sum \frac{1}{2} m_i v_{i/c}^2 + \sum \frac{1}{2} m_i v_c^2 + \sum \frac{1}{2} m_i 2 \vec{v}_{i/c} \cdot \vec{v}_c$

$= \sum \frac{1}{2} m_i v_{\frac{i}{c}}^2 + \frac{1}{2}(\sum m_i) v_c^2 + \left(\sum m_i \vec{v}_{\frac{i}{c}}\right) \cdot \vec{v}_c$

$= KE_{sys/cm} + \frac{1}{2} M v_c^2 + 0$

$KE_{sys} = KE_{sys/cm} + \frac{1}{2} M v_c^2$

or $\quad KE_{sys} = KE_{sys/cm} + \frac{P_c^2}{2M}$

Thus, the net KE of system of particle or body = to the KE of body about CM + KE of CM

For the system shown below-

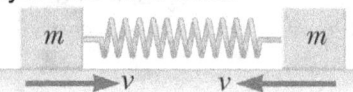

$K_{sys/CM} \neq 0$

but $\quad \frac{1}{2} M v c 2 = 0$

25. A SYSTEM OF TWO PARTICLES

Suppose the masses of the particles are equal to m_1 and m_2 and their velocities in an inertial reference frame attached to earth, are $\vec{v}_1$ and $\vec{v}_2$, respectively. Let us find the expressions defining their momenta and the total kinetic energy in the C-frame.

The momentum of the first particle in the C-system is
$P_{1/c} = m_1 \vec{v}_{1/c} = m_1(\vec{v}_1 - \vec{v}_{CM})$ ... (1)
where $\vec{v}_{CM}$ is the velocity of the centre of mass.
$\vec{v}_{CM} = \frac{m_1 v_1 + m_2 v_2}{m_1 + m_2}$

Substituting it in (1), we obtain
$p_{1/C} = m_1 \left(v_1 - \frac{m_1 v_1 + m_2 v_2}{m_1 + m_2}\right)$
$= \frac{m_1 m_2}{m_1 + m_2}(v_1 - v_2) = \mu(v_1 - v_2)$

In vector form,
or $\quad \vec{p}_{1/C} = \mu(\vec{v}_1 - \vec{v}_2)$
where $\mu = \frac{m_1 m_2}{m_1 + m_2}$, is called the **reduced mass** of the system

Similarly, the momentum of the second particle in the C frame is
$\vec{p}_{2/C} = \mu(\vec{v}_2 - \vec{v}_1)$

Thus, the momenta of the two particles in the C-frame are equal in magnitude and opposite in direction; the modulus of the momentum of each particle is
$\vec{p}_{1/C} = \mu v_{rel}$
where, $v_{rel} = |\vec{v}_1 - \vec{v}_2|$ is the velocity of one particle relative to another.

Finally, let us consider kinetic energy. The total kinetic energy of the two particles in the C-frame is

$K_{S/C} = K_1 + K_2 = \frac{\vec{p}^2}{2m_1} + \frac{\vec{p}^2}{2m_2}$

Since in accordance with equation $\mu = \frac{m_1 m_2}{m_1 + m_2}$ or $\frac{1}{m_2} +$ $\frac{1}{m_2} = \frac{1}{\mu}$, therefore

$K_{S/C} = \frac{p^2}{2\mu} = \frac{\mu v_{rel}^2}{2}$

If the particles interact, then total mechanical energy in the C frame is
$E = T + U$
where U is the potential energy of interaction of the given particles.

26. ANALYSIS OF SYSTEM OF TWO MASSES IN CM FRAME

There are numerous questions involving two masses which when observed from ground frame are either a little difficult to solve or lengthy in calculations.

A better option in such cases would be if we try to analyse the motion of the system from centre of mass (COM) frame.

Let us begin with a situation where we have two masses m_1 and m_2 moving with velocities $\vec{v}_1$ and $\vec{v}_2$ with respect to ground frame.

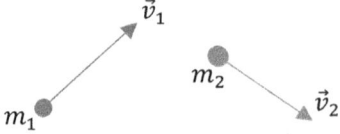

Hence, the velocity of COM becomes
$\vec{v}_C = \frac{m_1 \vec{v}_1 + m_2 \vec{v}_2}{m_1 + m_2}$

$\therefore$ The velocities of the masses m_1 and m_2 in the frame of COM becomes $\vec{v}_{1c}$ and $\vec{v}_{2c}$ where

$\vec{v}_{1c} = \vec{v}_1 - \vec{v}_c = \vec{v}_1 - \left(\frac{m_1 \vec{v}_1 + m_2 \vec{v}_2}{m_1 + m_2}\right)$

$= \frac{m_2(\vec{v}_1 - \vec{v}_2)}{m_1 + m_2}$

$\therefore \quad \vec{v}_{1c} = \frac{m_2}{m_1 + m_2}(\vec{v}_1 - \vec{v}_2)$

Similarly, $\vec{v}_{2c} = \frac{m_1}{m_1 + m_2}(\vec{v}_2 - \vec{v}_1)$

These equations just look like any ordinary equation but a close introspection shows a beautiful result as below:

The linear momentum of the masses m_1 and m_2 in COM frame becomes,

$\vec{p}_{1c} = m_1 \vec{v}_{1c} = \frac{m_1 m_2}{m_1 + m_2}(\vec{v}_1 - \vec{v}_2)$

$\vec{p}_{1c} = \mu(\vec{v}_1 - \vec{v}_2)$

where, $\mu = \frac{m_1 m_2}{m_1 + m_2}$ = reduced mass of the system

Similarly, $\vec{p}_{2c} = \mu(\vec{v}_2 - \vec{v}_1)$

Clearly note that $\vec{P}_{1c} = -\vec{P}_{2c}$, which means irrespective of the values of $\vec{v}_1$ and $\vec{v}_2$, whatever be their direction of motion, the magnitude of linear momentum of the masses with respect to centre of mass is equal in magnitude and opposite in direction.

Also, $p_{1c} = p_{2c} = \mu v_{rel}$
where, $v_{rel} = |\vec{v}_1 - \vec{v}_2|$
∴ The kinetic energy of the masses in COM frame
$K_{1c} = \frac{p_{1c}^2}{2m_1}$, $K_{2c} = \frac{p_{2c}^2}{2m_2}$
∴ Kinetic energy of the system in its COM frame becomes
$K_{cf} = K_{1c} + K_{2c} = \frac{1}{2}\left(\frac{p_{1c}^2}{m_1} + \frac{p_{2c}^2}{m_2}\right)$
but $p_{1c} = p_{2c} = \mu v_{rel}$
∴ $K_{cf} = \frac{1}{2}\left[\mu^2 v_{rel}^2 \left(\frac{1}{m_1} + \frac{1}{m_2}\right)\right] = \frac{1}{2}\left[\mu^2 v_{rel}^2 \times \frac{1}{\mu}\right] = \frac{1}{2}\mu v_{rel}^2$

Now, let us try to correlate the relation between kinetic energy in ground frame of system (K_{gf}) and kinetic energy in COM frame of system (K_{cf})

$K_{gf} = \frac{1}{2}m_1 v_1^2 + \frac{1}{2}m_2 v_2^2$
$= \frac{1}{2}m_1(\vec{v}_{1c} + \vec{v}_c)^2 + \frac{1}{2}m_2(\vec{v}_{2c} + \vec{v}_c)^2$
$= \frac{1}{2}m_1(v_{1c}^2 + v_c^2 + 2\vec{v}_{1c}\cdot\vec{v}_c) + \frac{1}{2}m_2(v_{2c}^2 + v_c^2 + 2\vec{v}_{2c}\cdot\vec{v}_c)$
∴ $K_{gf} = \left(\frac{1}{2}m_1 v_{1c}^2 + \frac{1}{2}m_2 v_{2c}^2\right) + \frac{1}{2}(m_1 + m_2)v_c^2$
$+ (m_1\vec{v}_{1c}\cdot\vec{v}_c + m_2\vec{v}_{2c}\cdot\vec{v}_c)$
where, $\frac{1}{2}m_1 v_{1c}^2 + \frac{1}{2}m_2 v_{2c}^2 = K_{cf}$
$\frac{1}{2}(m_1 + m_2)v_c^2$ = Kinetic energy of COM = K_{COM}
$m_1\vec{v}_{1c}\cdot\vec{v}_c + m_2\vec{v}_{2c}\cdot\vec{v}_c = (m_1\vec{v}_{1c} + m_2\vec{v}_{2c})\cdot\vec{v}_c = 0$
[since linear momentum of system in COM frame would be zero]
∴ $K_{gf} = K_{cf} + K_{COM}$
$= \frac{1}{2}\mu v_{rel}^2 + \frac{1}{2}(m_1 + m_2)v_c^2$
where v_c = velocity of COM in ground frame.
If instead of ground frame, we choose any other frame K_{COM} would represent the kinetic energy of COM in that frame.
Hence clearly, the COM frame is the frame of least kinetic energy.
Remember these results. They will help you in solving several questions, some of which have been shown below:

EXAMPLE 49. On a frictionless surface a block of mass $2m$ is projected towards an unstretched spring connected to a block of mass m. Find maximum compression in spring.

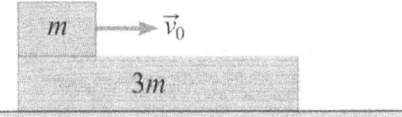

SOLUTION METHOD-1: In ground frame, on $(m + 2m)$ system, $F_{ext} = 0$, i.e., no change in linear momentum of system.
∴ $\Delta p = 0$
As soon as the $2m$ strikes spring, its velocity starts decreasing whereas of m starts increasing, due to which deformation starts increasing initially and reaches maximum till $2m$ travels faster than m. Hence at maximum compression, both attain same velocity.

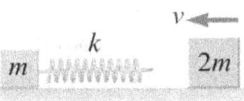

∴ $\Delta p = 0$
⇒ $2mv = (m + 2m)v_c$

or $v_c = \frac{2}{3}v$ ... (1)

Using work energy theorem
$W_{spring} = \Delta K$
⇒ $-\Delta U = \Delta K$
⇒ $-\left[\frac{1}{2}kx_{max}^2 - 0\right] = \frac{1}{2}(m + 2m)v_c^2 - \frac{1}{2}2mv^2$
⇒ $x_{max} = \sqrt{\frac{2m}{3k}}\,v$ [Using (1)]

METHOD-2: In COM frame,
$W_{spring} = \Delta K \Rightarrow -\Delta U = (K_{cf})_f - (K_{cf})_i$
Since, $(K_{cf})_f$ = final KE in COM frame
$= 0$ [∵ $v_{rel} = 0$]
∴ $-\left[\frac{1}{2}kx_{max}^2 - 0\right] = -\frac{1}{2}\left(\frac{m.2m}{3m}\right)v^2$
$\left[\because (K_{cf}) = \frac{1}{2}\mu v^2 = \frac{1}{2}\frac{m.2m}{m+2m}v_{rel}^2\right]$
⇒ $x_{max} = \sqrt{\frac{2m}{3k}}\,v$

EXAMPLE 50. A mass m is projected over a rough long plank of mass $3m$ kept on a smooth horizontal surface as shown.

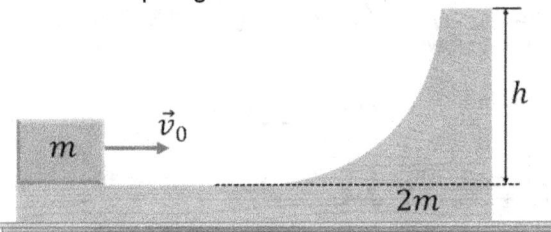

Find the work done by friction till relative slipping stops.
SOLUTION In COM frame,
$K_i = \frac{1}{2}\mu v_{rel}^2 = \frac{1}{2}\left(\frac{m.3m}{4m}\right)v_0^2 = \frac{1}{2}\left(\frac{3m}{4}\right)v_0^2$
$K_f = 0$ [∵ $v_{rel} = 0$]
∴ On applying work energy theorem, we get
$W_{friction} = \Delta K = 0 - \frac{1}{2}\left(\frac{3m}{4}\right)v_0^2$

EXAMPLE 51. All surfaces are smooth. Mass m is projected over $2m$ whose other end is almost vertical of total height h. The mass m is found to go to a height H above the top edge of mass $2m$. Find H.

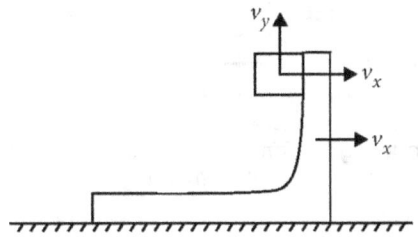

SOLUTION When the block m is about to leave $2m$, the situation will look like as shown in figure:

The horizontal component of velocity of m will be same as the velocity of $2m$, and thereafter, after leaving contact, its vertical component would change leaving horizontal component of both identical.
Hence at maximum height,

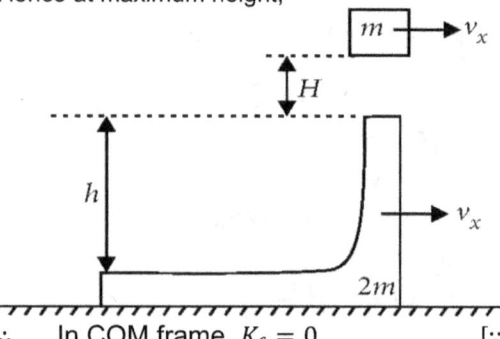

∴ In COM frame, $K_f = 0$ $\quad [\because \ v_{rel} = 0]$
∴ Applying work energy theorem,
$W_g = \Delta K$
$\Rightarrow \quad -\Delta U_g = (K_{cf})_f - (K_{cf})_i$
$\Rightarrow -[mg(h+H)] = -\frac{1}{2}\frac{2m}{3}\cdot v_0^2 \quad [\because \ \mu = \frac{m.2m}{m+2m} = \frac{2m}{3}]$
∴ $H = \frac{v_0^2}{3g} - h$

EXAMPLE 52. Two blocks are connected by an unstretched spring and mass m is given a velocity $v_0 = 24 \ ms^{-1}$ towards $5m$ as shown.

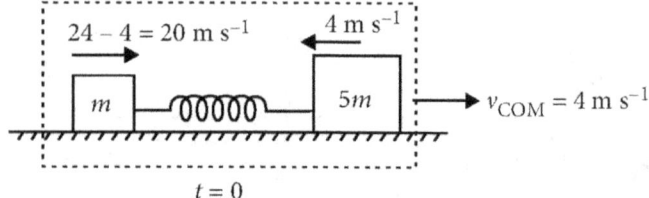

Find the maximum velocity of $5m$?
SOLUTION This is an example where the two masses oscillate about their COM frame (which moves with a constant velocity) with a time period,
$T = 2\pi\sqrt{\mu/k} = 2\pi\sqrt{5m/6k}$
Velocity of COM, $v_{COM} = \frac{mv_0}{m+5m} = \frac{v_0}{6} = 4 \ ms^{-1}$
∴ With respect to centre of mass, the given situation looks as shown in figure below:

[Diagram: at $t = 0$, $24 - 4 = 20$ m s⁻¹ for m, 4 m s⁻¹ for $5m$, $v_{COM} = 4$ m s⁻¹]

But after half oscillation, i.e. $t = T/2$ later
[Diagram: 20 m s⁻¹ for m, 4 m s⁻¹ for $5m$, $v_{COM} = 4$ m s⁻¹]

The velocity of any oscillating system is maximum at mean position only and here obviously the mean position is unstretched length.
But, $\vec{v}_{5m} = \vec{v}_{5mc} + \vec{v}_{COM}$

where $\vec{v}_{5m}$ = velocity of $5m$ in ground frame
$\vec{v}_{COM}$ = velocity of COM
$= 4ms^{-1}$ towards right
Now, if $\vec{A} = \vec{B} + \vec{C}$, then A is maximum only if $\vec{B}||\vec{C}$, hence
$|\vec{v}_{5m}|_{max}$ only if $\vec{v}_{5mc}||\vec{v}_{COM}$
∴ $(v_{5m})_{max} = (v_{5mc})_{max} + v_{COM} = 4 + 4 = 8ms^{-1}$

EXAMPLE 53. Two blocks A and B of masses m & $2m$ placed on smooth horizontal surface are connected with a light spring. The two blocks are given velocities as shown when spring is at natural length.

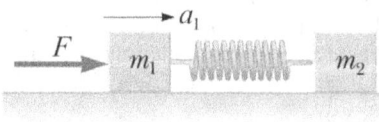

(i) Find velocity of centre of mass (b) maximum extension in the spring
SOLUTION Velocity of CM $v_{CM} = \frac{3mv - mv}{3m} = v$
In COM frame, initial momentum = 0
At the time of maximum elongation both the masses will be moving in same direction with same speed.
Initially relative velocity $v_{rel} = 3v$
By conservation of energy in CM frame,
Initial KE + Initial PE = Final KE + Final PE
$\frac{1}{2}\mu v_{rel}^2 + 0 = 0 + \frac{1}{2}kx^2$ (Initially, spring is un stretched, therefore Initial $PE = 0$ and finally $v_{rel} = 0$,
∴ Final KE = 0
$\Rightarrow \frac{1}{2}\frac{m \times 2m}{3m}(3v)^2 = \frac{1}{2}kx^2 \Rightarrow 3mv^2 = \frac{1}{2}kx^2 \Rightarrow x = v\sqrt{6m/k}$

EXAMPLE 54. Two blocks of masses $m_1 = 1kg$ and $m_2 = 2kg$ kept on smooth surface, are connected to each other through a light spring ($k = 100$ N/m) as shown in the figure.

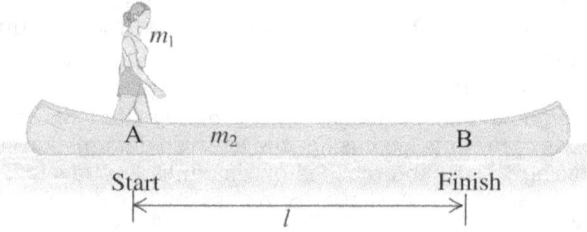

When we push mass m_1 with a force $F = 10 \ N$, mass m_1 is seen to move with acceleration $a_1 = 2 \ m/s^2$, what will be the acceleration of m_2?
SOLUTION From the FBD of M_1
$F - kx = m_1 a_1$ and from FBD of m_2, $kx = m_2 a_2$
∴ $a_2 = \frac{F - m_1 a_1}{m_2} = \frac{10N - (1 \ kg)(2m/s^2)}{2 \ kg} = 4 \ m/s^2$

EXAMPLE 55. If woman walks from A to B find displacement of woman and boat.

[Diagram: woman m_1 on boat m_2, from Start (A) to Finish (B), distance l]

SOLUTION Initially, woman boat system is at rest. Therefore, initial momentum of the women boat system is zero. As there is no external force on the system, therefore, by conservation of linear momentum, we have
Thus $\vec{P} = 0 \Rightarrow \vec{v}_{cm} = 0$
$\Rightarrow \Delta x_{cm} = 0$ (Shift of CM of system = 0)
$\Rightarrow m_1 \Delta x_1 + m_2 \Delta x_2 = 0$

Suppose when woman moves by distance l towards right with respect to boat, then boat moves by distance x toward right with respect to ground.

In this case shift in position of boat $\Delta x_2 = x$, and shift in position of woman $\Delta x_1 = l + x$

$\Rightarrow m_1(l + x) + m_2 x = 0$ or $\quad x = -\left(\dfrac{m_1}{m_1+m_2}\right)l$

here, negative sign shows that the shift of boat will be opposite to the shift of woman i.e., towards left.

EXAMPLE 56. A ball of mass m and radius R is placed inside a spherical shell of the same mass m and inner radius $2R$. The combination is at rest on a table top as shown in Fig. 1a. The ball is released, rolls back and forth inside, and finally comes to rest at the bottom, as in Fig. 1c. What will be the displacement d of the shell during this process?

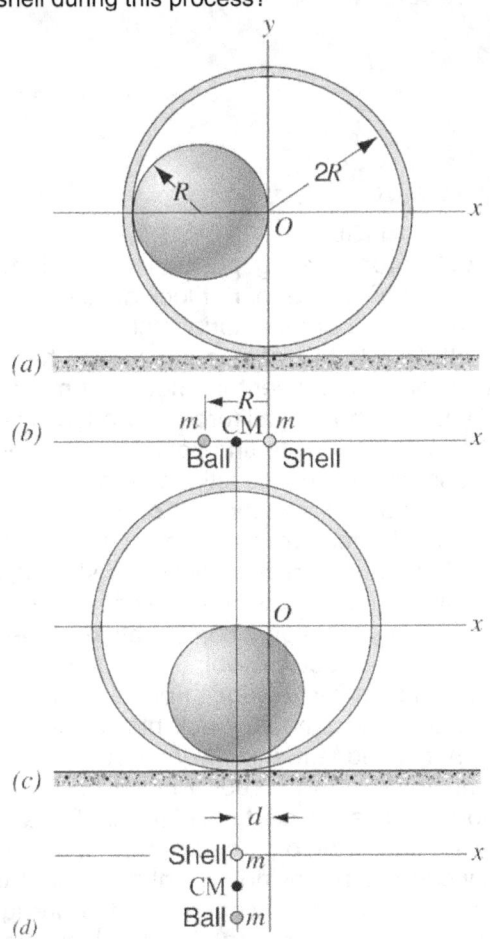

FIGURE 1. (a) A ball of radius R is released from this initial position and is free to roll inside a spherical shell of radius $2R$. (b) The centers of mass of the ball, the shell, and their combination. (c) The final state after the ball has come to rest. The shell has moved so that the horizontal coordinate of the center of mass of the system remains in place. (d) The centers of mass of the ball, the shell, and their combination.

APPROACH The only external forces acting on the ball–shell system are the downward force of gravity and the normal force exerted vertically upward by the table. Neither force has a horizontal component so that $\Sigma F_{ext,x} = 0$. Therefore, from equation $\Sigma F_{ext,x} = M a_{CM}$, the acceleration component $a_{CM,x}$ of the center of mass must also be zero. Thus, the horizontal position of the center of mass of the system must remain fixed, and the shell must move in such a way as to make sure that this happens.

We can represent both ball and shell by single particles of mass m, located at their respective centers. Fig 1b shows the system before the ball is released and Fig. 1d after the ball has come to rest at the bottom of the shell. We choose our origin to coincide with the initial position of the center of the shell. FIGURE 1b shows that, with respect to this origin, the center of mass of the ball–shell system is located a distance $\frac{1}{2}R$ to the left, halfway between the two particles. The shell must move to the left through this distance as the ball comes to rest.

SOLUTION If shift in position of CM of shell is $\Delta x_S = d$ along negative direction of x axis, then the displacement of the position of ball $\Delta x_b = R - d$, therefore

$m\Delta x_S + m\Delta x_b = 0 \quad \Rightarrow \quad m(-d) + m(R - d) = 0$
$\Rightarrow \quad d = R/2$

Fig. 1d shows that the displacement of the shell is given by, $d = (1/2)R$

DISCUSSION The ball is brought to rest by the frictional force that acts between it and the shell. Since it is an internal force of the system, therefore it will not affect the final location of the center of mass.

EXAMPLE 57. Inside a smooth spherical shell of the radius R a ball of the same mass is released from the shown position (Fig.1) Find the distance travelled by the shell on the horizontal floor when the ball comes to the lowest point of the shell.

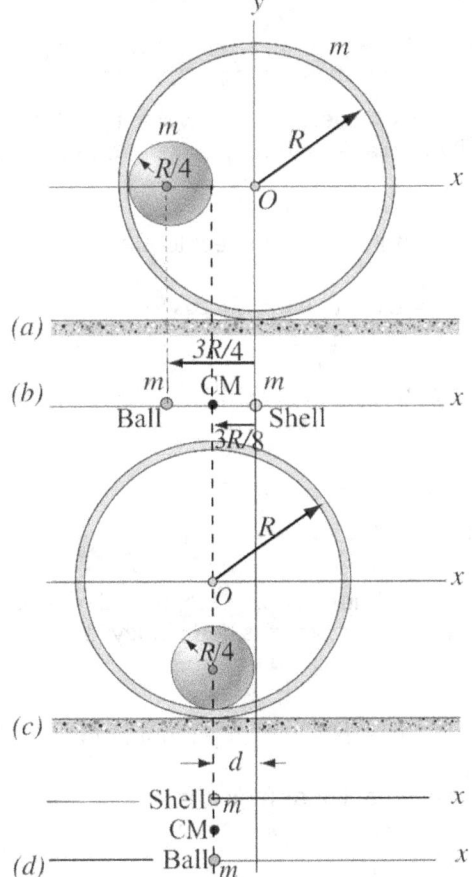

SOLUTION By conservation of linear momentum in horizontal direction, we have
$\Delta x_{CM} = 0$ (Shift in CM of the system = 0)
As radius of the ball is $R/4$, therefore its distance from the centre of spherical shell $= R - \frac{R}{4} = \frac{3R}{4}$
Suppose, when the ball reaches the lowest position of spherical shell, the spherical shell moves distance x towards left.
Therefore, the shift in the position of ball towards right with respect to ground $\Delta x_1 = (3R/4) - x$
Shift in the position of spherical shell $\Delta x_2 = -x$
$\therefore m\left(\frac{3R}{4} - x\right) - mx = 0$
or $x = 3R/8$

EXAMPLE 58. A frog sits at the end of a long board of length L. The board rests on a frictionless horizontal table. The frog wants to jump to the opposite end of the board. What is the minimum take-off speed i.e. relative to ground u that allows the frog to do the trick? The board and the frog have equal masses.

SOLUTION Since, there is no external force in horizontal direction, therefore, conserving linear momentum in horizontal direction, we get

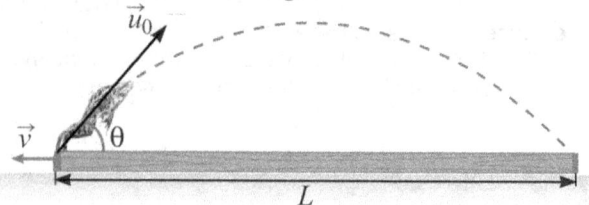

$mv = m(u\cos\theta)$, here u is the velocity of frog, and v is the velocity of plank with respect to ground as shown in figure.
$v = u\cos\theta$
Now velocity of frog with respect to plank
$v_{FP} = v_F - v_P$
or $v_{FP} = u\cos\theta + v = 2u\cos\theta$
Time of flight, $T = \frac{2u\sin\theta}{g}$,
$\therefore$ Range with respect to plank $= \frac{2u\sin\theta}{g} \times 2u\cos\theta = \frac{2u^2\sin 2\theta}{g}$
For maximum range, $\theta = \frac{\pi}{4}$
$\Rightarrow L = 2u^2/g$
or $u = \sqrt{Lg/2}$

EXAMPLE 59. When slipping stops, find total work done by friction assuming plank is sufficiently long.

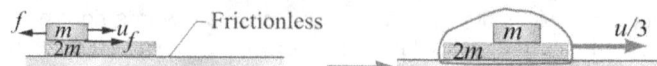

SOLUTION When slipping stops, both moves with same speed.
By conservation of linear momentum
$mu = 3mv$ or $v = u/3$

work done by friction = Change in $ME = \Delta KE$ (here we assumed gravitational PE at the plank level = 0)
$= K_f - K_i$
$= \frac{1}{2}2m(u/3)^2 + \frac{1}{2}m(u/3)^2 - \frac{1}{2}mu^2$
$= \frac{3mu^2}{18} - \frac{1}{2}mu^2 = -\frac{1}{3}mu^2$ Joules

EXAMPLE 60. Find the maximum height reached by small mass m as shown in FIGURE 1, assume all surfaces frictionless.

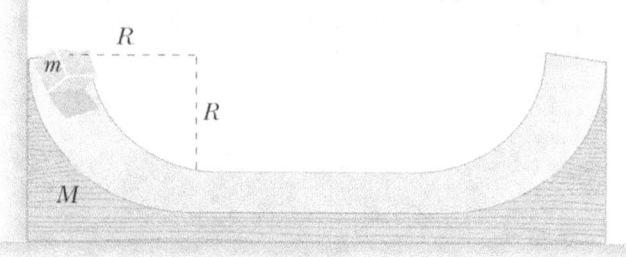

FIGURE 1

APPROACH From stating position of the block to its lowest position of circular part, block applies a reactional force on the wedge and its horizontal component will be towards left. Due to this component wedge has to move towards left but wall prevent its motion. It means wall is applying an equal horizontal force on wedge in right side. So, for the motion from starting point to bottom of circular part, there is an external reactional force of wall on the wedge in horizontal direction. So, we cannot apply conservation of linear momentum for the block-wedge system in horizontal direction between starting point to final position. Since there is no non-conservative force in this case, so, we can apply conservation of mechanical energy principle to get the velocity of smaller block at lowest position on the wedge. For next motion of the block on horizontal smooth track block applies normal reaction on the wedge in downward vertical direction, so, there is no force on the wedge in horizontal direction. When block reaches to next circular part it again applies a horizontal reactional force component in right side due to which wedge start moving in right direction. But in this case, there is no external force on the wedge block system in horizontal direction. Once block reaches its lower position the contact force applied by the wall in horizontal direction becomes zero and then for next motion we are free to apply conservation of linear momentum in horizontal direction. At the highest position of right circular track, block and wedge will have same horizontal velocity.

So, we first apply conservation of ME to determine the velocity of block at lowest position, then by conservation of linear momentum between lowest and highest position of right circular track, we can calculate the common horizontal velocity of block-wedge system. Now to find

the height oh smaller block on right circular track, we again apply conservation of mechanical energy.
SOLUTION It is given that, Mass of smaller block = m, mass of bigger block = M
Bigger block remains at rest till smaller block reaches at the bottom of circular part. After that bigger block also starts moving towards right side.
Velocity of smaller block at lowest point $u = \sqrt{2gR}$.

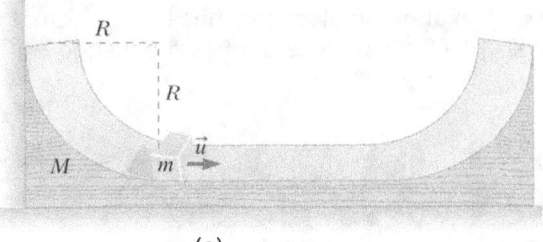

(a)

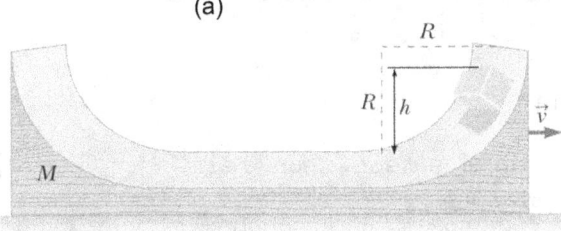

(b)
FIGURE 2

Suppose, smaller block reaches up to height $h (< R)$ at opposite side, as shown in figure, then at this moment both blocks will have same horizontal velocities. Let this velocity is v, the by conservation of linear momentum in horizontal direction, we have

$mu = (m + M)v$ or $v = \dfrac{m}{m+M}u$

By energy conservation, we have
Initial ME of the system when the block was at the lowest position = Final ME of the system when block is at its highest position
i.e., KE of the smaller block at lowest position = PE of the smaller block at highest position + KE of the smaller block + KE of bigger block
i.e., $\frac{1}{2}mu^2 = mgh + \frac{1}{2}mv^2 + \frac{1}{2}Mv^2$

$\qquad = mgh + \frac{1}{2}(m+M)v^2$

or $\quad mgh = \frac{1}{2}mu^2 - \frac{1}{2}(m+M)v^2$

$\qquad = \frac{1}{2}mu^2 - \frac{1}{2}(m+M) \times \left(\dfrac{m}{m+M}u\right)^2$

$\qquad = \frac{1}{2}mu^2\left(1 - \dfrac{m}{m+M}\right)$

or $\quad mgh = \frac{1}{2}mu^2\left(\dfrac{M}{m+M}\right) = \frac{1}{2}m(2gR)\left(\dfrac{M}{m+M}\right)$

or $\quad h = \left(\dfrac{M}{m+M}\right)R$

Special case: If $M = m$, then $h = R/2$

EXAMPLE 61. In the figure shown the wedge of mass M has a semicircular groove. A small ball of mass $m = M/2$ is released from A. It slides on the smooth circular track and starts climbing on the right face.

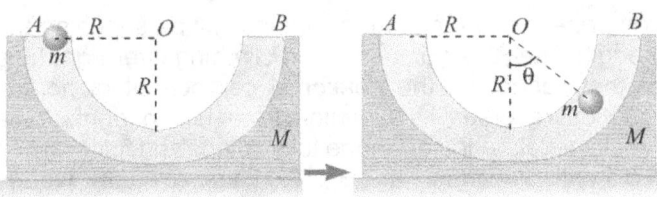

FIGURE 1

(I) Find the maximum value of θ which it can subtend with vertical and also find the distance displaced by wedge at this position.
(II) Find the maximum velocity of wedge during process of motion.
(I) APPROACH Initial momentum of the system, along X axis, is zero, therefore final momentum and relative velocity will also be zero.

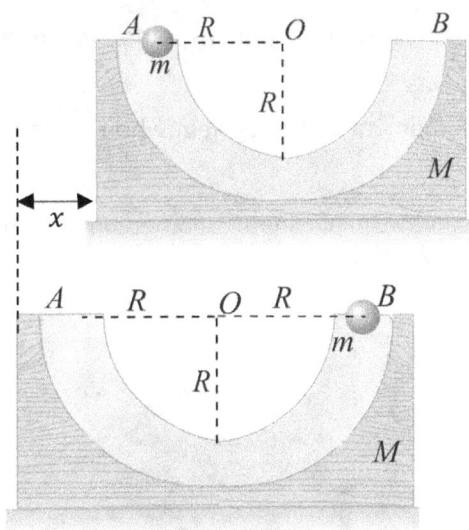

FIGURE 2

As all surfaces are smooth, and the only force acting in this case is conservative gravitational force, therefore mechanical energy of the system will also be conserved. As there is no external force in horizontal direction, therefore, horizontal shift in CM will be zero i.e.,
$\Delta x_{CM} = 0$
SOLUTION By conservation of mechanical energy, initial potential energy = final potential energy.
Hence, $\theta = 90°$
Let x is the displacement of M in negative X direction, therefore
$\Delta x_{CM} = 0$, gives
$m(2R - x) - Mx = 0$
or $\quad m(2R) = (M + m)x$
$x = \dfrac{2mR}{M+m} = \dfrac{2(M/2)R}{M+M/2}$
or $\quad x = \dfrac{2MR}{3M} = \dfrac{2R}{3}$

(II) APPROACH
Maximum velocity of wedge will be when the ball is at the lowest point in the wedge as till this point the horizontal

component of normal on the wedge will be speeding the wedge. After this point ball start climbing over right face of the wedge, so the horizontal component of normal reaction reverses its direction from left to right. As a result, deceleration of wedge takes place and finally when the ball reaches at B, the velocity of wedge instantaneously becomes zero. Now ball again starts sliding down but the horizontal component of normal reaction of ball on wedge is still in positive direction of x axis, therefore wedge starts moving towards positive direction of x axis. Again, at lowest position of the ball, the speed of the wedge will be maximum but this time it is towards positive direction of x axis. After this again retardation starts and the velocity of wedge instantaneously becomes zero when ball returns to A. This process remains continue and wedge keeps on oscillating.

SOLUTION Linear momentum of the system when the ball is at A,
$p_i = 0$
Linear momentum of the system when the ball is at its lowest position,
$p_f = -Mv + mu$
By conservation of linear momentum in horizontal direction, we have
$p_i = p_f$
$u = \frac{Mv}{m} = 2v$
$U_i + K_i = U_f + K_f$

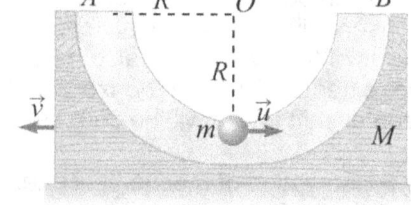

$mgR + 0 = 0 + \frac{1}{2}mu^2 + \frac{1}{2}Mv^2$ (considering, lowest position of the ball as reference potential level)
$2mgR = m(2v)^2 + Mv^2$
or $\quad 2 \times \frac{M}{2} \times gR = 4mv^2 + Mv^2$
or $\quad MgR = 4 \times \frac{M}{2} \times v^2 + Mv^2$
or $\quad MgR = 2Mv^2 + Mv^2$
or $\quad MgR = 3Mv^2$
or $\quad v = \sqrt{\frac{gR}{3}}$

27. CHECKPOINT 6

1. ••A small block of mass $m_1 = 0.500$ kg is released from rest at the top of a frictionless, curve-shaped wedge of mass $m_2 = 3.00$ kg, which sits on a frictionless, horizontal surface as shown in FIGURE P1A. When the block leaves the wedge, its velocity is measured to be 4.00 m/s to the right as shown in FIGURE P1b. (a) What is the velocity of the wedge after the block reaches the horizontal surface? (b) What is the height h of the wedge?

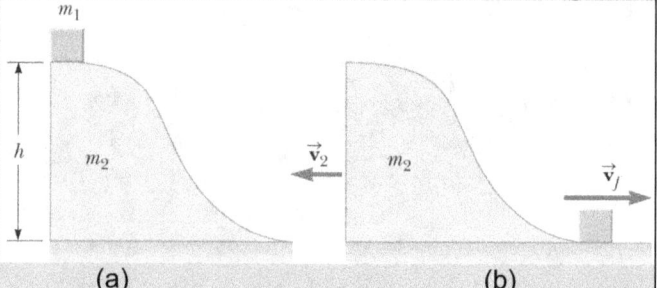

(a)   (b)
FIGURE P1

2. ••A man of mass m clings to a rope ladder suspended below a balloon of mass M (including the basket passenger); see Fig. P2. The balloon is stationary with respect to the ground. (a) If the man begins to climb the ladder at a speed v (with respect to the ladder), in what direction and with what speed (with respect to the Earth) will the balloon move? (b) What is the state of motion after the man stops climbing?

FIGURE P2

3. ••A cannon and a supply of cannonballs are inside a sealed railroad car of length L, as in Fig. P3. The cannon fires to the right; the car recoils to the left. The cannonballs remain in the car after hitting the far wall. (a) After all the cannonballs have been fired, what is the greatest distance the car can have moved from its original position? (b) What is the speed of the car after all the cannonballs have been fired?

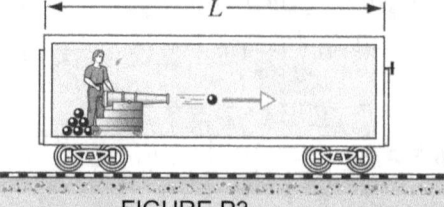

FIGURE P3

4. ••The three balls in the overhead view of Fig. P4 are identical. Balls 2 and 3 touch each other and are aligned perpendicular to the path of ball 1. The velocity of ball 1 has magnitude $v_0 = 10$ m/s and is directed at the contact point of balls 1 and 2.

After the collision, what are the (a) speed and (b) direction of the velocity of ball 2, the (c) speed and (d) direction of the velocity of ball 3, and the (e) speed and (f) direction of the velocity of ball 1? (Hint: With friction absent, each impulse is directed along the line connecting the centers of the colliding balls, normal to the colliding surfaces.)

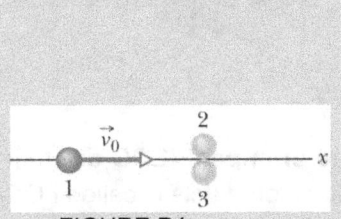
FIGURE P4

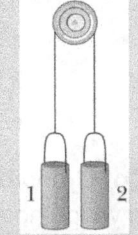
FIGURE P5

5. ••In Fig. P5, two identical containers of sugar are connected by a cord that passes over a frictionless pulley. The cord and pulley have negligible mass, each container and its sugar together have a mass of 500 g, the centers of the containers are separated by 50 mm, and the containers are held fixed at the same height. What is the horizontal distance between the center of container 1 and the center of mass of the two-container system (a) initially and (b) after 20 g of sugar is transferred from container 1 to container 2? After the transfer and after the containers are released, (c) in what direction and (d) at what acceleration magnitude does the center of mass move?

6. ••Two identical blocks of mass m, each are connected by a spring as shown in the FIGURE P6. At any instant of time $t = 0$, one block is given a velocity v_1 and other is given a velocity v_2 ($v_1 \gg v_2$) in the same direction simultaneously as shown in the figure. Find the maximum energy stored in the spring.

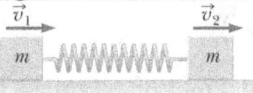
FIGURE P6

28. COLLISIONS

In physics, a **collision** is an isolated event in which two or more objects exert relatively strong forces on one another for a relatively short time. We usually assume the force between the colliding objects during the time of the collision is much stronger than any other forces exerted on either of them; this assumption is known as the ***impulse approximation***.

In the microscopic and submicroscopic world, our picture of a collision is different. When atoms collide, they don't "touch" each other: the atom doesn't have a definite spatial boundary, so there are no surfaces to make "contact." However, the collision model is still useful for atoms and subatomic particles whenever there is an interaction in which the forces are strong over a short time interval, so that there is a clear "before collision" and a clear "after collision."

28.1. CONSERVATION APPROACH IN A COLLISION

To analyze a collision with the conservation approach, we choose a system that includes the colliding objects. For objects (modeled as particles) to collide, at least one of the objects must be moving relative to the other(s). So, before the collision, the system has both momentum and kinetic energy. For the collisions we study, usually momentum is conserved (or nearly so), and kinetic energy may or may not be conserved.

28.1.1. CONSERVATION OF MOMENTUM DURING A COLLISION

If the system of colliding objects is closed (constant mass) and no net external force is exerted on the system, the system's momentum must be conserved. According to the impulse approximation, the forces exerted by the colliding objects on each other are often much stronger than any external forces; so, it may be acceptable to apply conservation of momentum to a system if the net external force is relatively weak. For example, if we are interested in the collision of an arrow and a tin can, we may ignore the gravitational force exerted by the Earth on the arrow–can system and apply the conservation of momentum principle to the system during the time of the collision. Of course, we must take gravity into account before and after the collision. Throughout our discussion, we assume *the total linear momentum of a system of colliding objects remains constant during the collision*.

$(\vec{p}_{tot})_i = (\vec{p}_{tot})_f$ ... (1)

Sometimes a more convenient way to express conservation of momentum is in terms of the system's center of mass. The total momentum of a system is equal to the momentum of the system's center of mass. So, conservation of momentum can be written as

$(\vec{p}_{CM})_i = (\vec{p}_{CM})_f$ ... (2)

From this equation, we can show that during a collision the center-of-mass velocity is conserved:

$$M(\vec{v}_{CM})_i = M(\vec{v}_{CM})_f$$

or $(\vec{v}_{CM})_i = (\vec{v}_{CM})_f$ ... (3)

A net external force is required to accelerate a system's center of mass. So, if there is no net external force (or if the net external force is negligible) during the collision, the center-of-mass velocity is not changed. The implications of conservation of momentum during a collision are summarized in the concept map shown below.

28.1.2. CONSERVATION OF KINETIC ENERGY DURING A COLLISION

You have probably noticed that when you drop a rubber ball it doesn't quite get back up to the original height after it bounces off the floor. Clearly the Earth–ball system loses some mechanical energy. Where does that energy go?

You may find the answer from another experience. If you have ever played a racket sport with a small rubber ball, you may have noticed that the ball gets very warm after it has been hit around for a few minutes. Some of the ball–wall system's kinetic energy has been converted into thermal energy. The system also loses energy to the surrounding air in the form of sound waves. During any collision, some or even all of the system's kinetic energy is converted into another form or is lost to the environment. In an **inelastic collision**, the system of colliding objects loses kinetic energy. So, the system's initial kinetic energy is greater than the kinetic energy after the collision:

$\vec{K}_{tot,i} > \vec{K}_{tot,f}$ (inelastic collision) ... (4)

The greatest loss of kinetic energy occurs when the colliding objects stick together in what is known as a **completely inelastic collision**. For example, a dart thrown into a dartboard is a completely inelastic collision. However, the objects do not need to be at rest after the collision. For example, if a sharpshooter shoots an apple placed on the head of a person with an arrow, the arrow will stick in the apple, and the apple and the arrow both will move together after the completely inelastic collision. An **elastic collision** is an ideal case in which the system of colliding objects conserves its total kinetic energy:

$\vec{K}_{tot,i} = \vec{K}_{tot,f}$ (Elastic collision) ... (5)

Although no actual collision is ever truly elastic, many collisions may be approximated as elastic.

So, although *linear momentum is always conserved* for a colliding system when external forces can be neglected, *kinetic energy is only conserved if the collision is elastic*. We apply conservation of momentum to both elastic and completely inelastic collisions, but we apply conservation of kinetic energy only to elastic collisions.

If after collision KE increases, then the collision is called a **super elastic collision**. For super elastic collision

$\vec{K}_{tot,i} < \vec{K}_{tot,f}$ (Elastic collision) ... (6)

The extra KE comes from elastic or some other PE already stored in the system.

☞ In collisions the two bodies try to occupy same space at same time.

28.2. MATHEMATICAL ANALYSIS OF COLLISION

When two bodies collide, they apply only normal contact force on each other through point(s) of contact.

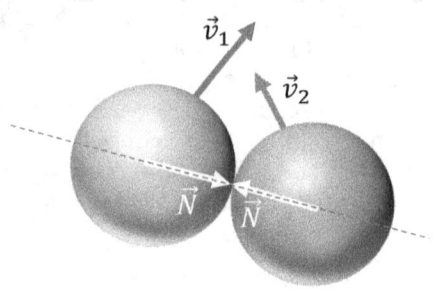

FIGURE 1

Line of Collision or line of impact (LOC or LOI): Line along which normal contact force acts is called LOC/LOI.
→ *LOC/LOI is independent of direction of velocity of colliding bodies.*

29. LAWS OF COLLISION

On the basis of directions, the collisions, between the bodies, can be classified into two categories-
1. Head-on collisions or one-dimensional collisions - where the velocity of each body just before impact is along the line of impact, and
2. Non-head-on collisions or oblique collisions or two-dimensional collisions - where the velocity of each body just before impact is not along the line of impact.

According to the coefficient of restitution (it is defined in next article), there are three special cases of any collision as written below:
1. Elastic collision
2. Inelastic collision
3. Super elastic collision

A detailed discussion of these collisions is given in next article.

29.1. HEAD ON (OR DIRECT) COLLISION (OR IMPACT)

29.1.1. CONSERVATION OF MOMENTUM

Suppose two particles of masses m_1 and m_2 have velocities $\vec{u}_1$ and $\vec{u}_2$ just before collision. After collision their velocities are $\vec{v}_1$ and $\vec{v}_2$ respectively. By conservation of linear momentum,

Before collision $m_1 \xrightarrow{u_1}$ $m_2 \xrightarrow{u_2}$

After collision v_1 v_2

Line of collision(LOC)
FIGURE 1

we have
$$m_1\vec{v}_1 + m_2\vec{v}_2 = m_1\vec{u}_1 + m_2\vec{u}_2$$
(This equation is generally written along the line of collision)
If both balls are moving in same direction, then we can

write above equation in scalar form, as
$$m_1 v_1 + m_2 v_2 = m_1 u_1 + m_2 u_2 \quad \ldots (1)$$
or $\quad m_2(v_2 - u_2) = m_1(u_1 - v_1) \quad \ldots (1a)$

29.1.2. CONSERVATION OF KINETIC ENERGY

29.1.2.1. ELASTIC COLLISION

A collision is said to be elastic if kinetic energy is also conserved along with the linear momentum. There is no loss or transformation of kinetic energy.

$K_f = K_i$

$\frac{1}{2}m_1 v_1^2 + \frac{1}{2}m_2 v_2^2 = \frac{1}{2}m_1 u_1^2 + \frac{1}{2}m_2 u_2^2$

or $\quad m_1 v_1^2 + m_2 v_2^2 = m_1 u_1^2 + m_2 u_2^2 \quad \ldots (2)$

or $\quad m_2(v_2^2 - u_2^2) = m_1(u_1^2 - v_1^2) \quad \ldots (2a)$

29.1.3. CALCULATION OF FINAL VELOCITIES AFTER ELASTIC COLLISION

On dividing Eq. 2a by Eq.1a, we get-

$\frac{m_2(v_2^2 - u_2^2)}{m_2(v_2 - u_2)} = \frac{m_1(u_1^2 - v_1^2)}{m_1(u_1 - v_1)}$

or $\quad \frac{m_2(v_2 - u_2)(v_2 + u_2)}{m_2(v_2 - u_2)} = \frac{m_1(u_1 - v_1)(u_1 + v_1)}{m_1(u_1 - v_1)}$

or $\quad v_2 + u_2 = u_1 + v_1$

or $\quad v_2 = u_1 + v_1 - u_2 \quad \ldots (3)$

Substituting this value of v_2 in Eq. 1a, we get-

$m_2(u_1 + v_1 - u_2 - u_2) = m_1(u_1 - v_1)$

or $\quad m_2 u_1 + m_2 v_1 - 2m_2 u_2 = m_1 u_1 - m_1 v_1$

or $\quad v_1(m_2 + m_1) = (m_1 - m_2)u_1 + 2m_2 u_2$

or $\quad v_1 = \left(\frac{m_1 - m_2}{m_1 + m_2}\right)u_1 + \left(\frac{2m_2}{m_1 + m_2}\right)u_2 \quad \ldots (4)$

Substituting this value of v_1 in Eq. 3, we get-

$v_2 = u_1 + \left(\frac{m_1 - m_2}{m_1 + m_2}\right)u_1 + \left(\frac{2m_2}{m_1 + m_2}\right)u_2 - u_2$

or $\quad v_2 = \frac{m_1 u_1 + m_2 u_1 + m_1 u_1 - m_2 u_1 + 2m_2 u_2 - m_1 u_2 - m_2 u_2}{m_1 + m_2}$

or $\quad v_2 = \frac{m_1 u_1 + m_1 u_1 + m_2 u_2 - m_1 u_2}{m_1 + m_2}$

or $\quad v_2 = \frac{2m_1 u_1 + (m_2 - m_1)u_2}{m_1 + m_2}$

or $\quad v_2 = \left(\frac{m_2 - m_1}{m_1 + m_2}\right)u_2 + \left(\frac{2m_1}{m_1 + m_2}\right)u_1 \quad \ldots (5)$

SPECIAL CASES

1. If $m_1 = m_2$, then
   From Eq. (4), we have-
   $v_1 = \left(\frac{m_1 - m_1}{m_1 + m_1}\right)u_1 + \left(\frac{2m_1}{m_1 + m_1}\right)u_2$

or $\quad v_1 = u_2 \quad \ldots (A)$

Substituting this value of v_1, in Eq. 3, we get-

$v_2 = u_1 + u_2 - u_2$

or $\quad v_2 = u_1 \quad \ldots (B)$

From Eq. (A), and (B), it is clear that velocities get interchanged after elastic collision between two bodies of equal masses.

29.1.3.1. INELASTIC COLLISION

Those collision in which momentum are conserved, kinetic energy are not conserved.

Loss in KE is $\Delta K = \left(\frac{1}{2}m_1 u_1^2 + \frac{1}{2}m_2 u_2^2\right) - \left(\frac{1}{2}m_1 v_1^2 + \frac{1}{2}m_2 v_2^2\right)$

If after collision, both bodies stick together, then loss in kinetic energy

$\Delta K = \frac{1}{2}m_1 u_1^2 + \frac{1}{2}m_2 u_2^2 - \frac{1}{2}(m_1 + m_2)v^2$,

here v is the common velocity of particles after collision. By conservation of LM, we have

$(m_1 + m_2)v = m_1 u_1 + m_2 u_2$

or $\quad v = \frac{m_1 u_1 + m_2 u_2}{m_1 + m_2}$

$\therefore \quad \Delta K = \frac{1}{2}(m_1 u_1^2 + m_2 u_2^2) - \frac{1}{2}\left(\frac{m_1 u_1 + m_2 u_2}{(m_1 + m_2)}\right)^2$

$= \frac{1}{2}\frac{m_1 m_2}{(m_1 + m_2)}(u_1 - u_2)^2$

☞ An inelastic collision doesn't have to be *completely* inelastic. Inelastic collisions include many situations in which the bodies do *not* stick.

☞ **Inelastic vs. Perfectly Inelastic Collisions** If the colliding particles stick together, the collision is perfectly inelastic. If they bounce off each other (and kinetic energy is not conserved), the collision is inelastic.

29.1.3.2. SUPER-ELASTIC COLLISION

In this type of collision, LM remains conserved but KE of system increases. The extra KE comes from elastic PE or from some other PE already stored in the system.

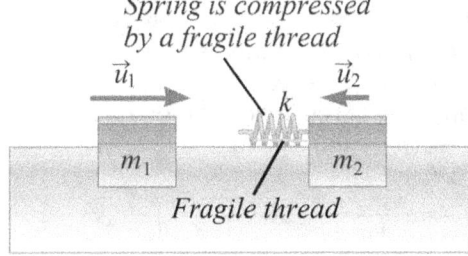

FIGURE 1

In above figure, two gliders are set in motion on a horizontal air track. A spring of force constant k is attached to the front face of the second glider. It is already compressed with a fragile thread. The first glider, of mass m_1, is moving to the right with speed u_1, and the second glider, of mass m_2, is moving to the left with speed u_2. When m_1 collides with the compressed spring attached to m_2, the system comes instantaneously into rest (maximum deformation position), then the fragile thread get break down and the gliders move apart again with more kinetic energies. This extra kinetic energy is provided by potential energy stored in compressed spring.

So, in super elastic collision, linear momentum remains conserved whereas KE increases.

29.2. NEWTON'S LAW FOR COLLISION

29.2.1. CONCEPT OF COEFFICIENT OF RESTITUTION

When two bodies collide head–on, the ratio of their relative velocity after collision and their relative velocity before collision is called the coefficient of restitution. Thus, $\frac{v_2-v_1}{u_2-u_1} = -e$ (coefficient of restitution)

This Law is valid even when momentum is not conserved:
1. For perfectly elastic collision $e=1$, for inelastic collision $0 < e < 1$
for perfectly inelastic collision (Bodies will stick together) $e=0$ and for super elastic collision $e > 1$
2. Newton's law of restitution can also be written as-
$$v_2 - v_1 = -e(u_2 - u_1)$$
or $\quad v_2 - v_1 = e(u_1 - u_2) \quad \ldots (6)$
i.e., relative velocity of receding = $e \times$ relative velocity of approach before impact

29.2.2. CALCULATION OF VELOCITIES AFTER DIRECT (OR HEAD ON) IMPACT BY USING NEWTON'S LAW OF RESTITUTION

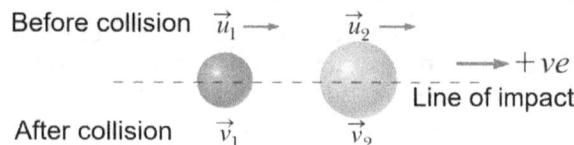

By Conservation of linear momentum,
$m_1 u_1 + m_2 u_2 = m_1 v_1 + m_2 v_2 \quad \ldots (1)$
By Newton's law of restitution, we have
$v_2 - v_1 = -e(u_2 - u_1) \quad \ldots (2)$
From (1) & (2)
$$\left. \begin{array}{l} v_1 = \frac{(m_1-em_2)u_1}{m_1+m_2} + \frac{(1+e)m_2 u_2}{m_1+m_2} \\ v_2 = \frac{(m_2-em_1)u_2}{m_1+m_2} + \frac{(1+e)m_1 u_1}{m_1+m_2} \end{array} \right\} \quad \ldots (3)$$

SPECIAL RESULT If $m_1 \gg m_2$, (i.e. m_1 is like earth or wall or any other heavy object) then from Eq. (3), we have
$v_2 = \frac{-em_1 u_2}{m_1} + \frac{(1+e)m_1 u_1}{m_1}$ and $\quad v_1 \approx u_1$
$$v_2 = (1+e)u_1 - eu_2$$

(i) If $m_1 \gg m_2$ and $u_1 = 0$, then $v_1 = 0$ but $v_2 = -eu_2$

(ii) If $m_1 \gg m_2$ and $u_2 = 0$, then $v_1 = u_1$ but $v_2 = (1+e)u_1$

1. For perfectly Inelastic (or plastic) collision, $e=0$

$\therefore \quad v_1 = v_2 = \frac{m_1 u_1 + m_2 u_2}{m_1 + m_2} = v$ (say)

Loss in KE, $\Delta K = \left(\frac{1}{2}m_1 u_1^2 + \frac{1}{2}m_2 v_2^2\right) - \frac{1}{2}(m_1+m_2)v^2$
substituting the value of v, we get
$\Delta K = \frac{1}{2}\left(\frac{m_1 m_2}{m_1+m_2}\right)(u_1 - u_2)^2 \quad \ldots (3)$

2. For elastic collision, $e=1$,
$\therefore \quad v_1 = \frac{(m_1-m_2)u_1}{m_1+m_2} + \frac{2m_2 u_2}{m_1+m_2}$
and $\quad v_2 = \frac{(m_2-m_1)u_2}{(m_1+m_2)} + \frac{2m_1 u_1}{m_1+m_2}$
$v_2 = 2u_1 - eu_2$

(i) If $m_1 \gg m_2$ and $u_1 = 0$, then $v_1 = 0$ but $v_2 = -u_2$

(ii) If $m_1 \gg m_2$ and $u_2 = 0$, then $v_2 = 2u_1$

SPECIAL CASES
CASE I: If both bodies have the same mass, then $m_1 = m_2 = m$ (say), then from (1) and (2), we get
$v_1 = u_2$ and $v_2 = u_1$
Thus, the velocities get interchanged in elastic collision between particles having same masses.

CASE II: If one of the bodies say m_2 is initially at rest, i.e., $u_2 = 0$. In this case, we have
$v_1 = 0, v_2 = u_1$

POINTS TO BE NOTICE
(i) The maximum transfer of energy occurs if $m_1 = m_2$,
(ii) If K_i and K_f are the initial and final kinetic energies of mass m_1, the fractional decrease in its kinetic energy is given by
$\frac{K_i - K_f}{K_i} = 1 - \frac{v_1^2}{u_1^2}$
Further, if $m_2 = nm_1$ and $u_2 = 0$, then
$\frac{K_i - K_f}{K_i} = \frac{4n}{(1+n)^2}$
(iii) If $m_1 \gg m_2$, then
$v_1 \approx u_1$
$v_2 = 2u_1 - u_2$
(iii) In an elastic collision of two equal masses, their kinetic energies are exchanged.

30. CHECKPOINT 7

1. • Find which ball will move when first ball is released and collide with the second ball in Newton's cradle shown below. All balls are of same mass and all collisions are elastic.

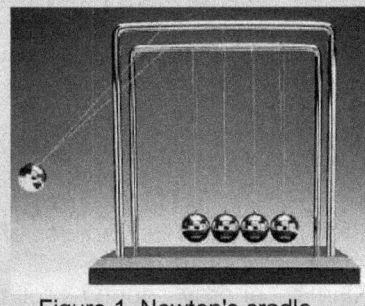

Figure 1. Newton's cradle

EXAMPLE 62. *A moving particle of mass m, makes a head–on collision with a particle of mass 2m, which is initially at rest. Show that the colliding particle loses (8/9) of its energy after collision.*

SOLUTION Let u be the initial velocity of particle of mass m and v its velocity after the collision. Let V be the velocity of particle of mass $2m$ after the collision.
From the principle of conservation of linear momentum, we have
$mu = mv + (2m)V$
or $\quad\quad u - v = 2V \quad\quad\quad\quad …(1)$
The conservation of kinetic energy gives
$\frac{1}{2}mu^2 = \frac{1}{2}mv^2 + \frac{1}{2}(2m)V^2$
or $\quad u^2 - v^2 = 2V^2$
or $\quad (u-v)(u+v) = 2V^2 \quad\quad …(2)$
Using Eq. (1) in Eq. (2) we have
$2V(u+v) = 2V^2$ or $u + v = V$
or $\quad\quad 2(u+v) = 2V \quad\quad\quad …(3)$
Comparing (1) and (3) we get
$\quad\quad\quad u - v = 2(u+v)$
or $\quad\quad u - v = 2(u+v)$
or $\quad\quad v = -\frac{u}{3} \quad\quad\quad\quad …(4)$
Now, initial kinetic energy of the colliding mass is
$K_i = \frac{1}{2}mu^2$
Final kinetic energy, $\quad\quad K_f = \frac{1}{2}mv^2$
Loss in kinetic energy is $\Delta K = K_i - K_f = \frac{1}{2}mu^2 - \frac{1}{2}mv^2$
∴ Fractional loss $= \frac{\Delta K}{K_i}$
$= \frac{\frac{1}{2}mu^2 - \frac{1}{2}mv^2}{\frac{1}{2}mu^2}$
$= \frac{u^2 - v^2}{u^2} = 1 - \left(\frac{v}{u}\right)^2$
$= 1 - \left(-\frac{1}{3}\right)^2 \quad\quad \left(\because v = -\frac{u}{3}\right)$
$= \frac{8}{9}$

EXAMPLE 63. *A neutron moving at a speed v undergoes a head-on elastic collision with a nucleus of mass number A at rest. Prove that the ratio of the kinetic energies of the neutron after and before collision is $\left(\frac{A-1}{A+1}\right)^2$.*

SOLUTION Mass of neutron $(m_1) = 1$ unit. Mass of nucleus $(m_2) = A$ units. Here $u_1 = u$ and $u_2 = 0$. Therefore, the velocity of the neutron after the collision is
$v_1 = \left(\frac{m_1 - m_2}{m_1 + m_2}\right)u = \left(\frac{1-A}{1+A}\right)u$
KE of neutron after collision $= \frac{1}{2}m_1v_1^2 = \frac{1}{2} \times 1 \times \left(\frac{1-A}{1+A}\right)^2 u^2$
KE of neutron before collision $= \frac{1}{2}m_1u^2 = \frac{1}{2} \times 1 \times u^2 = \frac{1}{2}u^2$. Their ratio is $\left(\frac{1-A}{1+A}\right)^2$.

EXAMPLE 64. *In a collinear collision, a particle with an initial speed v_0 strikes a stationary particle of the same mass. If the final total kinetic energy is 50% greater than the original kinetic energy, the magnitude of the relative velocity between the two particles, after collision, is*
[JEE Main 2018]

(A) $\frac{v_0}{4}$ (B) $\sqrt{2}v_0$ (C) $\frac{v_0}{2}$ (D) $\frac{v_0}{\sqrt{2}}$

APPROACH Since, after collision kinetic energy is increasing, therefore it is an example of super elastic collision. To find relative velocity $|v_1 - v_2|$ after collision, we have to calculate velocities v_1 and v_2. To get v_1 and v_2, we need two equations. One equation can be obtained by applying principle of conservation of linear momentum, i.e.,
$m_1u_1 + m_2u_2 = m_1v_1 + m_2v_2 \quad\quad …(1)$
and other equation can be obtained, by using the given relation
$K_f = K_i + \frac{50}{100}K_i$
i.e., $\quad K_f = K_i + \frac{1}{2}K_i \quad\quad\quad …(2)$

Before collision $\quad \vec{u}_1 = v_0 \quad\quad \vec{u}_2 = 0$
$m_1 = m \quad\quad m \quad\quad$ Line of impact

After collision $\quad \vec{v}_1 \quad\quad \vec{v}_2$

SOLUTION It is given that, $m_1 = m_2 = m$ (say), $u_1 = v_0$, $u_2 = 0$, therefore from (1), we get
$mv_0 + m.0 = mv_1 + mv_2$
or $\quad\quad v_1 + v_2 = v_0 \quad\quad\quad …(3)$
From equation (2), we get
$\frac{1}{2}mv_1^2 + \frac{1}{2}mv_2^2 = \frac{1}{2}mv_0^2 + \frac{1}{2}\left(\frac{1}{2}mv_0^2\right)$
or $\quad v_1^2 + v_2^2 = v_0^2 + \frac{1}{2}v_0^2$
or $\quad v_1^2 + v_2^2 = \frac{3}{2}v_0^2 \quad\quad …(4)$
Now, $(v_1 - v_2)^2 = (v_1 + v_2)^2 - 4v_1v_2$
or $\quad v_1^2 + v_2^2 - 2v_1v_2 = v_0^2 - 4v_1v_2$
or $\quad \frac{3}{2}v_0^2 - 2v_1v_2 = v_0^2 - 4v_1v_2$
or $\quad 2v_1v_2 = -\frac{1}{2}v_0^2 \quad\quad …(5)$
Now again,
$(v_1 - v_2)^2 = (v_1 + v_2)^2 - 4v_1v_2 = (v_1 + v_2)^2 - 2(2v_1v_2)$
Using (3) and (5) in above equation we get
$(v_1 - v_2)^2 = v_0^2 - 2\left(-\frac{1}{2}v_0^2\right) = 2v_0^2$
or $\quad v_1 - v_2 = \sqrt{2}v_0$
So, option (B) is correct.

31. COLLISION OF A BALL WITH GROUND

Now, let's examine what happens in the limit where one of the two colliding objects is the ground (for all intents and purposes, infinitely massive) and the other one is a ball. We can see from equation (2) that if the ground does not move when the ball bounces, $u_1 = v_1 = 0$, and we can write for the speed of the ball:
$v_2 = -eu_2$ ... (4)
If we release the ball from some height, h_0, we know that it reaches a speed of $v_0 = \sqrt{2gh_0}$ immediately before it collides with the ground. If the collision is elastic, the speed of the ball just after the collision is the same, $v_1 = v_0 = \sqrt{2gh_0}$, and it bounces back to the same height from which it was released. If the collision is totally inelastic, as in the case of a ball of putty that falls to the ground and then just stays there, the final speed is zero. For all cases in between, we can find the coefficient of restitution from the height h_1 that the ball returns to:
$$v^2 = u^2 - 2gh$$
$\Rightarrow \quad 0 = v_1^2 - 2gh_1$
$\Rightarrow \quad h_1 = \frac{v_1^2}{2g} = \frac{e^2 v_0^2}{2g} = e^2 h_0$
or $\quad e^2 = \frac{h_1}{h_0} = \frac{v_1^2}{v_0^2}$ ... (5)

In general, we can state (without proof) that the kinetic energy loss in partially inelastic collisions is
$\Delta K = K_0 - K_1 = \frac{1}{2}\frac{m_1 m_2}{m_1+m_2}(1-e^2)(u_1-u_2)^2$ ... (6)

In the limit $e \to 1$, we obtain $\Delta K = 0$, i.e., no loss in kinetic energy, as required for elastic collisions. In addition, in the limit $e \to 0$, equation (6) matches the energy release for totally inelastic collisions already shown in equation (3).

31.1. GENERALIZATION

31.1.1. CALCULATION OF HEIGHT AFTER n^{th} IMPACT

From equation (5), the height covered after first collision with ground is
$h_1 = e^2 h_0$
Similarly, after 2nd impact, the height covered
$h_2 = e^2 h_1 = e^4 h_0$
after third impact, $h_3 = e^2 h_2 = e^6 h_0$
and after n^{th} impact, $h_n = e^{2n} h_0$

31.1.2. CALCULATION OF VELOCITY AFTER n^{th} IMPACT

Again from (5), we have
Velocity after first impact, $v_1 = ev_0$
Velocity after second impact, $v_2 = ev_1 = e^2 v_0$

Similarly, velocity after n^{th} impact, $v_n = e^n v_0$

31.1.3. TOTAL DISTANCE COVERED BY BALL

Total distance covered by the ball, to come to rest, is given by
$h = h_0 + 2(h_1 + h_2 + h_3 + \cdots)$
$= h_0 + 2e^2 h_0(1 + e^2 + e^4 + \cdots)$
$= h_0 + 2e^2 h_0 \left(\frac{1}{1-e^2}\right) = h_0 \left[\frac{1+e^2}{1-e^2}\right]$

EXAMPLE 65. Two particles of mass m_1, m_2 moving with initial velocity u_1 and u_2 collide head-on. Find minimum Kinetic energy during collision. Thus, prove that maximum kinetic energy is lost in perfectly inelastic collision

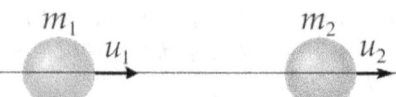

SOLUTION In C-frame initial kinetic energy of system if $\frac{1}{2}\mu(v_2 - v_1)^2$ where $\mu = \frac{m_1 m_2}{m_1+m_2}$. During collision at the instant of maximum deformation we get minimum kinetic energy in C-frame as they attain same velocity thus no relative velocity. When system have minimum kinetic energy in C-frame it also has minimum kinetic energy in ground frame as velocity of CM is constant.
$K_G = \frac{1}{2}\mu v_{rel}^2 + \frac{1}{2}m_s v_c^2$ at maximum deformation. Thus, minimum kinetic energy during collision is $\frac{1}{2}(m_1 + m_2)v_c^2$, where $v_c = \frac{(m_1 u_1 + m_2 u_2)}{m_1+m_2}$

In perfectly inelastic collision, $v_{rel} = 0$, $\therefore$ Final kinetic energy is $K_f = \frac{1}{2}m_s v_c^2$ which is minimum as discussed above.

EXAMPLE 66. BALLISTIC PENDULUM A ballistic pendulum (Fig. 1) is a device that was used to measure the speeds of bullets before electronic timing devices were available. It consists of a large block of wood of mass M, hanging from two long pairs of cords. A bullet of mass m is fired into the block, and the block bullet combination swings upward, its center of mass rising a vertical distance h before the pendulum comes momentarily to rest at the end of its arc.

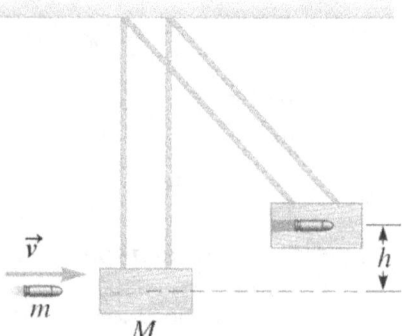

FIGURE 1 Ballistic Pendulum

Take the mass of the block to be $M = 5.4$ kg and the mass

of the bullet to be $m = 9.5$ g. (a) What is the initial speed of the bullet if the block rises to a height of $h = 6.3$ cm? (b) What fraction of the initial kinetic energy is lost in this collision?

APPROACH (a) Let us divide the problem into two parts: (1) The bullet moving with speed v_i enters the block and comes to rest relative to the block, after which the bullet block combination moves with a common speed v_f. We assume this happens very quickly. (2) The combination, now moving with speed v_f, swings upward until it comes to rest. Part 1 is an example of a completely inelastic collision, in which the two colliding objects stick together after the collision. So, in this part we can apply conservation of linear momentum but not conservation of mechanical energy (as the collision is inelastic). In second part we can apply either conservation of mechanical energy (as gravitational force is conservative in nature) or work energy theorem.

SOLUTION In part 1, by conservation of linear momentum, we have

$$mv_i = (m+M)v_f \quad \ldots (1)$$

For part 2, we are going to apply work energy theorem (although, you can also use conservation of mechanical energy). The net work on the block bullet combination is that done by gravity: $W_{net} = W_g = -(m+M)gh$, and as it swings upward and comes to rest the change in the kinetic energy of the combination is $\Delta K = 0 - \frac{1}{2}(m+M)v_f^2$. The work–energy theorem, $W_{net} = \Delta K$, then gives

$$-(m+M)gh = -\frac{1}{2}(m+M)v_f^2 \quad \ldots (2)$$

Substituting the value of v_f from (1) in (2), we get

$$-(m+M)gh = -\frac{1}{2}(m+M)\left(\frac{mv_i}{m+M}\right)^2,$$

Solving for v_i, we find

$$v_i = \left(\frac{M+m}{m}\right)\sqrt{2gh}$$
$$= \left(\frac{5.4\ kg + 0.0095\ kg}{0.0095\ kg}\right)\sqrt{(2)(9.8\ m/s^2)(0.063\ m)}$$
$$= 630\ m/s$$

We can look at the ballistic pendulum as a kind of transformer, exchanging the high speed of a light object (the bullet) for the low— and thus more easily measurable—speed of a massive object (the block).

(b) We can write the final kinetic energy as

$$K_f = \frac{1}{2}(m+M)v_f^2 = \frac{1}{2}(m+M)\left(\frac{mv_i}{m+M}\right)^2$$
$$= \frac{1}{2}mv_i^2\left(\frac{m}{m+M}\right)$$

The ratio between the initial and final kinetic energies is

$$\frac{K_f}{K_i} = \frac{m}{m+M} = \frac{9.5\ g}{9.5\ g + 5.4\ kg} = 0.0018$$

Only 0.18% of the initial kinetic energy remains after the collision. The remaining 99.82% is stored inside the pendulum as internal energy (perhaps in part as a temperature increase) or transferred to the environment—for example, as heat or sound waves.

COMMENT: Note that during the collision, momentum is conserved but not kinetic energy (the collision is totally inelastic); and that during the swinging motion, the total mechanical energy is conserved, but not momentum (the swinging motion proceeds under the influence of the "external" forces of gravity and the tensions in the wires).

> As the strings are very light and we are applying momentum and energy concept in analyzing the motion of ballistic pendulum, therefore for mathematical purpose we can replace both strings by just a single string. In next problems we use just single string not two.

EXAMPLE 67. The ballistic pendulum shown in FIGURE 1a consists of a stationary block of wood of mass M suspended by a wire of negligible mass. A bullet of mass m is fired into the block, and the block (with the bullet in it) swings to a maximum height of h above the initial position (see part 1b of the drawing). Find the speed with which the bullet is fired, assuming that air resistance is negligible.

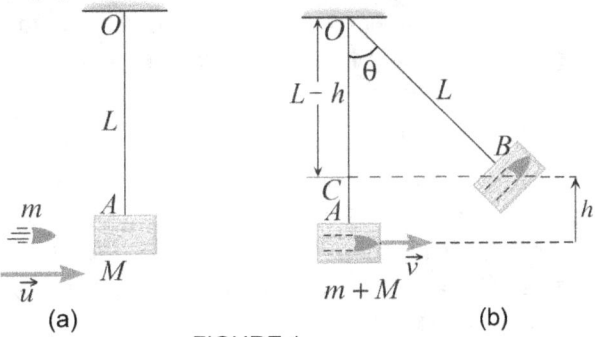

FIGURE 1

APPROACH The physics of the ballistic pendulum can be divided into two parts. The first is the completely inelastic collision between the bullet and the block. The total linear momentum of the system (block plus bullet) is conserved during the collision, because the suspension wire supports the system's weight, which means that the sum of the external forces acting on the system is nearly zero. The second part of the physics is the resulting motion of the block and bullet as they swing upward after the collision. As the system swings upward, the principle of conservation of mechanical energy applies, since nonconservative forces do no work. The tension force in the wire does no work because it acts perpendicular to the motion. Since air resistance is negligible, we can ignore the work it does. The conservation principles for linear momentum and mechanical energy provide the basis for our solution.

SOLUTION STEP 1 Completely Inelastic Collision

Just after the bullet collides with it, the block (with the bullet in it) has a speed v. Since linear momentum is conserved, the total momentum of the block–bullet system after the collision is the same as it is before the collision:

$$\underbrace{(m+M)v}_{\text{Total momentum after collision}} = \underbrace{mu + 0}_{\text{Total momentum before collision}}$$

Note that the block is at rest before the collision, so its initial momentum is zero.

From above, we get

$$u = \frac{m+M}{m}v \quad \ldots (1)$$

STEP 2 Conservation of Mechanical Energy

The speed v immediately after the collision can be obtained from the maximum height h to which the system swings, by using the principle of conservation of mechanical energy:

$$\underbrace{(m+M)gh}_{\text{Total mechanical energy at top of swing, all potential}} = \underbrace{\tfrac{1}{2}(m+M)v^2}_{\text{Total mechanical energy at bottom of swing, all kinetic}}$$

This result can be solved for v,

$$v = \sqrt{2gh} \qquad \ldots (2)$$

Calculation of speed of bullet

Substituting the value of v, from (2) in (1), we get

$$u = \underbrace{\tfrac{m+M}{m}v}_{\text{Step 1}} = \underbrace{\tfrac{m+M}{m}\sqrt{2gh}}_{\text{Step 2}} \qquad \ldots (3)$$

From equation (3), we can calculate speed of bullet 'u'.

32. MISCONCEPTION

1. In applying the mechanical energy conservation principle, it is tempting to say that the total potential energy at the top of the swing is equal to the total kinetic energy of the bullet before it strikes the block $\left[(m+M)gh = \tfrac{1}{2}mu^2\right]$ and solve directly for u. This is incorrect, however, because the collision between the bullet and the block is inelastic, so that some of the bullet's initial kinetic energy is dissipated during the collision (due to friction and structural damage to the block and bullet).

2. When an object collides inelastically with a stationary object having much greater mass (like wall, ground or trunk of tree), nearly all of the first object's kinetic energy is lost in the form of sound and thermal energy. In this problem, the bullet and wood become hotter as mechanical energy is converted to internal energy.

EXAMPLE 68. A bullet of mass m moving with a horizontal velocity u strikes a stationary block of mass M suspended by a string of length L. The bullet gets embedded in the block. What is the maximum angle made by the string after impact.

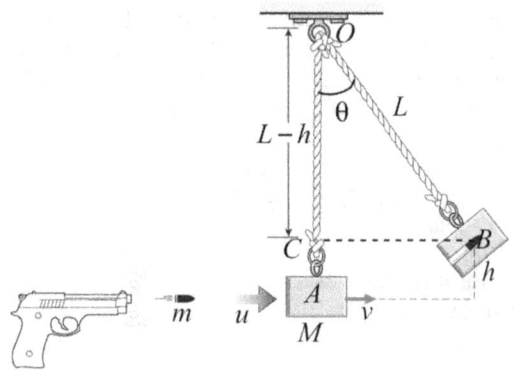

APPROACH The approach is similar as discussed in ballistic pendulum.

SOLUTION Let v be the combined velocity of the bullet-block system just after collision, then by conservation of linear momentum.

$$mu = (m+M)v \qquad \ldots (1)$$

After collision the KE of the bullet-block system gets converted into potential energy.

$$\therefore \quad \tfrac{1}{2}(m+M)v^2 = (m+M)gh$$

or $\qquad v = \sqrt{2gh}$

Putting this value of v in equation (1), we get

$$mu = (m+M)\sqrt{2gh}$$

or $\qquad h = \dfrac{u^2}{2g}\left[\dfrac{m}{m+M}\right]^2$

Form the figure, $\cos\theta = \dfrac{OC}{OB} = \dfrac{L-h}{L} = 1 - \dfrac{h}{L}$

$$\therefore \quad \theta = \cos^{-1}\left[1 - \dfrac{u^2}{2gL}\left(\dfrac{m}{m+M}\right)^2\right]$$

☞ Here we cannot apply principle of conservation of energy before and after collision between m and M, because collision between m and M is inelastic.

33. OBLIQUE OR INDIRECT COLLISION

☞ Use this concept only when line of collision is known or can be known.

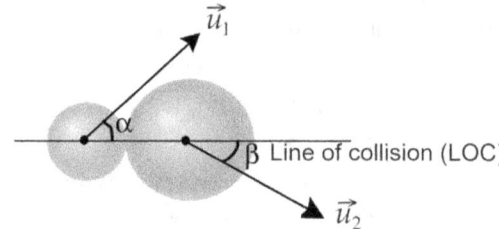

Case (i) If one of them was at rest then after collision the body which was at rest will move in direction of line of collision.

Case (ii) If two moving bodies collide obliquely but one of them continues to move in same direction as was before colliding then that direction is direction of line of collision. Since there is no force perpendicular to LOC, therefore LM of individual bodies or linear velocities along perpendicular to LOC remains unchanged.

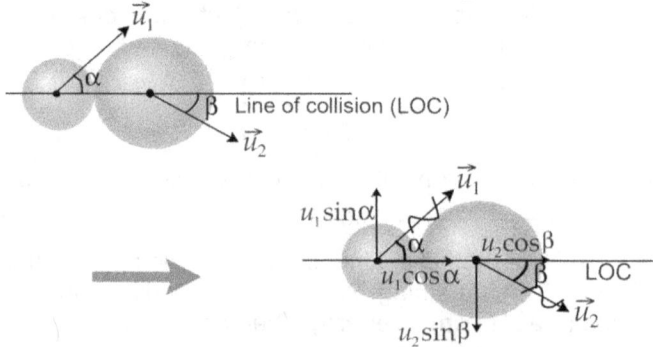

So, $u_1 \sin\alpha$ & $u_2 \sin\beta$ will remain unchanged.

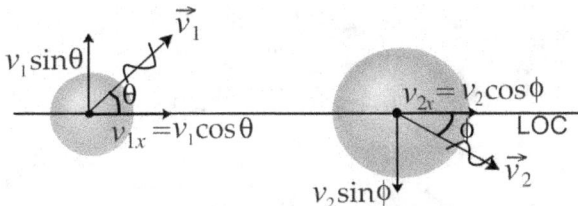

If $\vec{v}_1$, $\vec{v}_2$ are the velocities of the bodies 1 and 2 respectively after collision. Then their components will be-
$v_1 \sin \theta$, $v_2 \sin \theta \to$ perpendicular to LOC
$v_{1x} = v_1 \cos \theta$, $v_{2x} = v_2 \sin \theta \to$ along LOC
Since perpendicular components remains unchanged, therefore
$v_1 \sin \theta = u_1 \sin \alpha$, $\quad v_2 \sin \phi = u_2 \sin \beta$
By conservation of LM, along LOC, we have
$m_1 u_1 \cos \alpha + m_2 u_2 \cos \beta = m_1 v_{1x} + m_2 v_{2x}$
and from figure, we can write
$v_1 = \sqrt{(v_{1x})^2 + (u_1 \sin \alpha)^2}$; $\quad \tan \theta = \dfrac{u_1 \sin \alpha}{v_{1x}}$
$v_2 = \sqrt{(v_{2x})^2 + (u_1 \sin \beta)^2}$; $\quad \tan \phi = \dfrac{u_2 \sin \beta}{v_{2x}}$
$\theta + \phi \to$ angle of divergence (see Figure).
By using above equations, we can find all unknowns asked in such problems.

EXAMPLE 69. A ball of mass 2 kg, moving with a velocity of 3 m/sec, impinges on a ball of mass 4 kg moving with a velocity of 1 m/sec. The velocities of two balls are parallel and inclined at 30° to the line joining their centers at the instant of impact. If the coefficient of restitution be 0.5, find

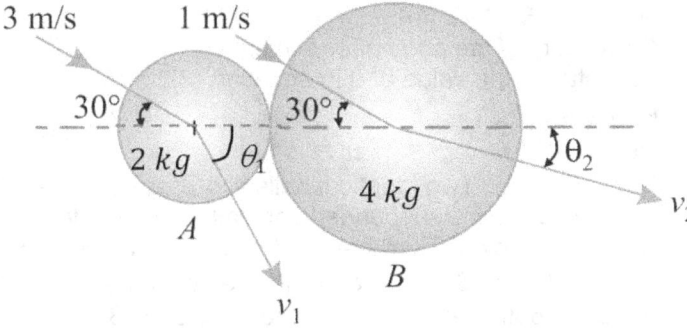

a) direction, in which the 4 kg ball will move after impact;
b) velocity of the 4 kg ball after impact;
c) direction, in which the 2 kg ball will move after impact; and
d) velocity of the 2 kg ball after impact.

APPROACH Since the colliding bodies are not moving along the line of impact, therefore it is an example of oblique or indirect collision. So, we will apply the principle discussed in indirect collision.

SOLUTION Given: Mass of first ball (M_1) = 2 kg; Initial velocity of first ball (u_1) = 3 m/s; Mass of second ball (m_2) = 4 kg; Initial velocity of second ball (u_2) = 1 m/s; Angle, which initial velocity of first ball makes with the line of impact (α_1) = 30°; Angle, which initial velocity of second ball makes with the line of impact (α_2) = 30° and coefficient of restitution (e) = 0.5

(a) Direction, in which the 4 kg ball will move after the impact
Let θ_1 = Angle, which the 2 kg ball makes with the line of impact,
θ_2 = Angle, which the 4 kg ball makes with the line of impact,
v_1 = Velocity of the 2 kg ball after impact, and
v_2 = Velocity of the 4 kg ball after impact,
We know that the components of velocities, perpendicular to the line of impact, remain unchanged before and after impact.
$\therefore \quad v_1 \sin \theta_1 = u_1 \sin \alpha_1 = 3 \sin 30° = 3 \times 0.5$
or $\quad v_1 \sin \theta_1 = 1.5$ ...(1)
Similarly, $v_2 \sin \theta_2 = u_2 \sin \alpha_2 = 1 \sin 30° = 1 \times 0.5$
or $\quad v_2 \sin \theta_2 = 0.5$ ...(2)
We also know from the law of conservation of momentum
$m_1 u_1 \cos \alpha_1 + m_2 u_2 \cos \alpha_2 = m_1 v_1 \cos \theta_1 + m_2 v_2 \cos \theta_2$
$(2 \times 3 \cos 30°) + (4 \times 1 \times \cos 30°)$
$= 2 v_1 \cos \theta_1 + 4 v_2 \cos \theta_2$
$(6 \times 0.866) + (4 \times 0.866) = 2 v_1 \cos \theta_1 + 4 v_2 \cos \theta_2$
$8.66 = 2 v_1 \cos \theta_1 + 4 v_2 \cos \theta_2$
$\therefore \quad v_1 \cos \theta_1 + 2 v_2 \cos \theta_2 = 4.33$ ...(3)
We know from the law of collision of elastic bodies that
$v_2 \cos \theta_2 - v_1 \cos \theta_1 = e(u_1 \cos \alpha_1 - u_2 \cos \alpha_2)$
$= 0.5 (3 \cos 30° - 1 \cos 30°)$
$= 0.5(3 \times 0.866 - 1 \times 0.866)$
$v_2 \cos \theta_2 - v_1 \cos \theta_1 = 0.866$ ...(4)
Adding equations (iii) and (iv),
$3 v_2 \cos \theta_2 = 5.196$
or $v_2 \cos \theta_2 = 1.732$ ...(5)
Dividing equation (2) by (5),
$\tan \theta_2 = \dfrac{0.5}{1.732} = 0.2887$
or $\quad \theta_2 = 16.1°$ **Ans.**

(b) Velocity of the 4 kg ball after impact
Substituting the value of θ_2 in equation (2),
$v_2 \sin 16.1° = 0.5$
$\therefore \quad v_2 = \dfrac{0.5}{\sin 16.1°} = \dfrac{0.5}{0.2773} = 1.803$ m/s **Ans.**

(c) Direction, in which the 2 kg ball will move after impact
Substituting the values of θ_2 and v_2 in equation (4),
$1.803 \cos 16.1° - v_1 \cos \theta_1 = 0.866$
or $v_1 \cos \theta_1 = 1.803 \cos 16.1° - 0.866$
$= (1.803 \times 0.9608) - 0.866 = 0.866$ ...(6)
Dividing equation (1) by (6)
$\tan \theta_1 = 1.5/0.866 = 1.732 \quad$ or $\quad \theta_1 = 60°$ **Ans.**

(d) Velocity of 2 kg ball after impact
Now substituting the value of θ_1 in equation (1), we get
$v_1 \sin 60° = 1.5$
$\therefore v_1 = 1.5/\sin 60° = 1.5/0.866 = 1.732$ m/s **Ans.**

34. INDIRECT IMPACT OF A BODY WITH A FIXED PLANE

Consider a body having an indirect impact on a fixed

plane as shown in Fig. 1.

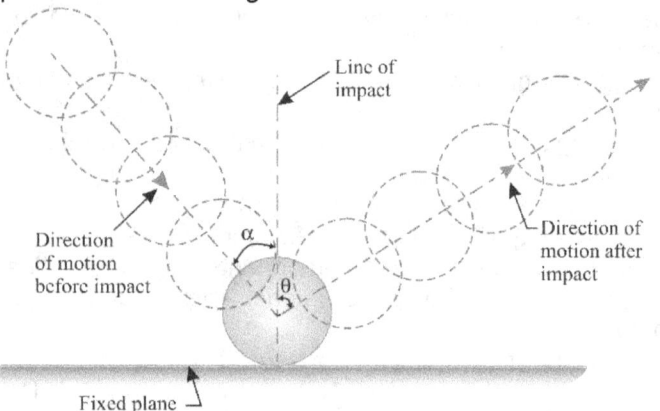

FIGURE 1

Let u = Initial velocity of the body,
v = Final velocity of the body,
α = Angle, which the initial velocity of the body makes with the line of impact,
θ = Angle which the final velocity of the body makes with the line of impact, and
e = Coefficient of restitution.

A little consideration will show, that the component of u, along the line of impact will cause the direct 'impact' of the body with the fixed plane. The other component of u (i.e. along the perpendicular to the line of impact) will not affect the phenomenon of impact and will be equal to the other component of v (i.e., along the perpendicular to the line of impact).

We know that velocity of approach = $u \cos \alpha$
and velocity of separation = $v \cos \theta$
The Newton's Law of Collision of Elastic Bodies also holds good for this impact i.e.,
$v \cos \theta = eu \cos \alpha$ ... (1)

Notes:
1. In this impact also, we cannot apply the principle of conservation of linear momentum because the fixed plane has infinite mass.
2. The components of initial and final velocities at right angles to the line of impact are same i.e.
$u \sin \alpha = v \sin \theta$ ... (2)
For elastic collision, equation (1) gives
$u \cos \alpha = v \cos \theta$ ... (3)
Dividing, (2) by (4), we get
$\tan \alpha = \tan \theta$
i.e., $\alpha = \theta$

EXAMPLE 70. A ball, moving with a velocity of 4 m/s, impinges on a fixed plane at an angle of 30°. If the coefficient of restitution is 0.5, find,

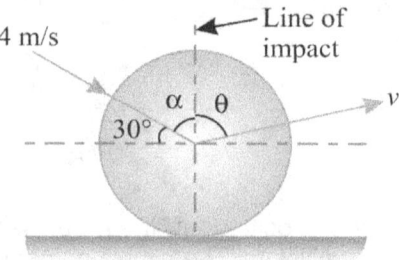

FIGURE 1

(a) direction of the body after impact, and
(b) velocity of the body after impact.

SOLUTION Given: Initial velocity of the body $(u) = 4$ m/s; Angle, which the initial velocity of the body makes with the line of impact $(\alpha) = 90° - 30° = 60°$ and coefficient of restitution $(e) = 0.5$.

(a) Direction of the body after impact
Let θ = Angle, which the final velocity makes with the line of impact, and v = Final velocity of the body after impact.
By conservation of linear momentum parallel to the plane,
$u \sin \alpha = v \sin \theta$
$\therefore v \sin \theta = u \sin \alpha = 4 \sin 60°$
or $v \sin \theta = 4 \times 0.866 = 3.464$ ... (1)
We also know from the law of collision of elastic bodies that
$v \cos \theta = e \times u \cos \alpha$
$= 0.5 \times 4 \cos 60° = 2 \times 0.5 = 1$ ... (2)
Dividing equation (1) by (2), we get
$\dfrac{v \sin \theta}{v \cos \theta} = \dfrac{3.464}{1}$
or $\tan \theta = 3.464$
or $\theta = 73.9°$ Ans.

(b) velocity of the body after impact
Substituting the value of θ in equation (2),
$v \cos 73.9° = 1$
or $v = \dfrac{1}{\cos 73.9°} = \dfrac{1}{0.2773} = 3.6$ m/s Ans.

EXAMPLE 71. Two identical balls A & B each of mass 2 kg & radius R are suspended vertically from inextensible strings as shown. Third ball C of mass 1 kg & radius $r = (\sqrt{2} - 1)R$ falls & hits A & B symmetrically with 10 m/s. Speed of both A & B just after the collision is 3 m/s.

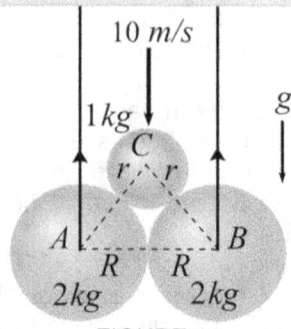

FIGURE 1

(a) Speed of C just after collision is
(A) 2 m/s (B) $2\sqrt{2}$ m/s

(C) 5 m/s (D) $(\sqrt{2} - 1)$ m/s
(b) Impulse provided by each string during collision is
(A) $6\sqrt{2}$ N sec. (B) 12 N sec.
(C) $3\sqrt{2}$ N sec. (D) 6 N sec.
(c) The value of coefficient of restitution is
(A) $\frac{1}{4}$ (B) $\frac{1}{\sqrt{2}}$
(C) $(\sqrt{2} - 1)$ (D) $\frac{1}{2}$

SOLUTION (a) As the balls A & B are constrained to move horizontally, if $\vec{J}$ be the impulse imparted by ball 'C' to each of A & B, then A and B will also apply equal and opposite impulses to ball C (See Figure 2a and b).
Since, horizontal components of impulses on C get cancelled, therefore the impulse received by ball C from both A and B would be $2J \sin \theta$ (in vertically upward direction).
From FIGURE 2b, we have
$\cos \theta = \frac{AL}{AC} = \frac{R}{R+r} = \frac{R}{R+(\sqrt{2}-1)R} = \frac{R}{\sqrt{2}R} = \frac{1}{\sqrt{2}}$
or $\theta = \frac{\pi}{4}$

Each of ball A & B received impulse $\vec{J}$ as shown, but moves horizontally as its vertical components gets balanced by impulse imparted to ball A & B by the respective strings, therefore, for balls A and B in horizontal direction

Ball A	Ball B
$J \cos \theta = m_A v_A$ (along $-ve$ direction of X axis) $\Rightarrow J = \frac{m_A v_A}{\cos \theta}$ $= \sqrt{2} m_A v_A$ $= \sqrt{2} \times 2 \times 3$ $= 6\sqrt{2}$ kg.m/s	$J \cos \theta = m_B v_B$ (along $+ve$ direction of X axis) $\Rightarrow J = \frac{m_B v_B}{\cos \theta}$ $= \sqrt{2} m_B v_B$ $= \sqrt{2} \times 2 \times 3$ $= 6\sqrt{2}$ kg.m/s

$\therefore \quad J = \sqrt{2} m_A v_A = \sqrt{2} m_B v_B$

Now, for **ball C**, if its final velocity is v'_C in downward direction, then
$m_C v'_C = m_C v_C - 2J \sin \frac{\pi}{4} = m_C v_C - \frac{2J}{\sqrt{2}}$
$= m_C v_C - \frac{2\sqrt{2} m_A v_A}{\sqrt{2}} == m_C v_C - 2 m_A v_A$
or $v'_C = v_C - 2 \frac{m_A}{m_C} v_A = 10 - 2 \times \left(\frac{2}{1}\right) \times 3 = -2$ m/s
(−ve sign indicates that v'_C is directed upwards)

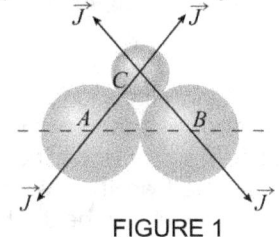
FIGURE 1

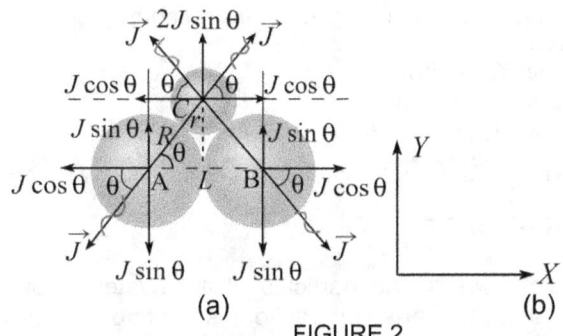

(a) (b)
FIGURE 2

(b) Impulse provided by each string
$J \sin \theta = J \sin \frac{\pi}{4} = 6\sqrt{2} \times \frac{1}{\sqrt{2}} = 6$ N sec.

(c) If before and after collision velocities of ball C, are v_1 and v'_1 respectively; v_2, v'_2 are that of ball A along the LOC, then by Newton's law of restitution along LOC, we have

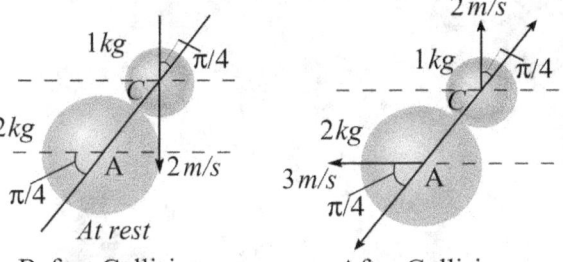

Before Collision After Collision
FIGURE 3

$v'_2 - v'_1 = -e(v_2 - v_1)$
or $3 \cos \frac{\pi}{4} - \left(-2 \cos \frac{\pi}{4}\right) = -e \left(0 - 10 \cos \frac{\pi}{4}\right)$
$e = \frac{(3/\sqrt{2}) + (2/\sqrt{2})}{10/\sqrt{2}} = \frac{5}{10} = \frac{1}{2}$

35. IMPORTANT POINTS

1. In a collision we consider situation before and after the collision. These terms refer to conditions when interactive force between particles effectively becomes zero. The duration of collision is negligible as compared to the time for which we observe the event.
2. In a collision effect of external non-impulsive (e.g. gravity, friction) forces can be neglected because they are very small compared to impulsive forces & time of collision is very small.
3. When two bodies collide, they exert force on each other through point of contact, perpendicular to the plane of contact. The direction of force of interaction is line of collision.
4. In case of collision if the external impulsive forces are not acting in a direction, the total momentum of system in that direction remains conserved
   i.e. $m_1 \vec{u} + m_2 \vec{u}_2 = m_1 \vec{v}_1 + m_2 \vec{v}_2$
5. After collision, only the components of velocity along line of collision changes, the $\perp$ components of velocity remain unaffected.
6. According to initial velocities and line of collision, collisions are divided into two types:
   (i) head on (ii) oblique

7. According to conservation of mechanical energy collision is of three types:
8. (i) elastic: $KE_f = KE_i$
9. (ii) inelastic: $KE_f < KE_i$
10. (iii) Super elastic: $KE_f > KE_i$

36. EXPLOSIONS AND CRASH-LANDINGS

An explosion, where the particles of the system move apart after a brief, intense interaction, is the opposite of a collision. The explosive forces, which could be from an expanding spring or from expanding hot gases, are *internal* forces. If the system is isolated, its total momentum during the explosion will be conserved.

Figure 1 These exploding rockets produce a spectacular display of bright sparks in the night sky.

The rockets shown in Figure 1 rise high into the sky. As they start to fall, they send out showers of chemical packages, each of which explodes to produce a brilliant sphere of burning chemicals. Material flies out in all directions to create a spectacular effect.

Does an explosion create momentum out of nothing? The important point to note here is that the burning material spreads out equally in all directions. Each tiny spark has momentum, but for every spark, there is another moving in the opposite direction, i.e. with opposite momentum. Since momentum is a vector quantity, the total amount of momentum created is zero.

At the same time, kinetic energy is created in an explosion. Burning material flies outwards; its kinetic energy has come from the chemical potential energy stored in the chemical materials before they burn.

More fireworks

A roman candle fires a jet of burning material up into the sky. This is another type of explosion, but it doesn't send material in all directions. The firework tube directs the material upwards. Has momentum been created out of nothing here?

Again, the answer is no. The chemicals have momentum upwards, but at the same time, the roman candle pushes downwards on the Earth. An equal amount of downwards momentum is given to the Earth. Of course, the Earth is massive, and we don't notice the tiny change in its velocity which results.

Down to Earth

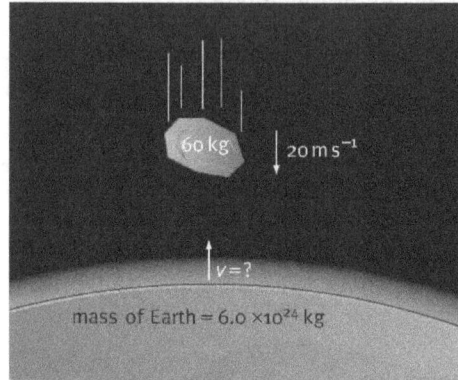

FIGURE 2 The rock and Earth gain momentum in opposite directions.

If you push a large rock over a cliff, its speed increases as it falls. Where does its momentum come from? And when it lands, where does its momentum disappear to? The rock falls because of the pull of the Earth's gravity on it. This force is its weight and it makes the rock accelerate towards the Earth. Its weight does work and the rock gains kinetic energy. It gains momentum downwards. Something must be gaining an equal amount of momentum in the opposite (upward) direction. It is the Earth, which starts to move upwards as the rock falls downwards. The mass of the Earth is so great that its change in velocity is small – far too small to be noticeable. When the rock hits the ground, its momentum becomes zero. At the same instant, the Earth also stops moving upwards. The rock's momentum cancels out the Earth's momentum. At all times during the rock's fall and crash-landing, momentum has been conserved.

If a rock of mass 60 kg is falling towards the Earth at a speed of 20 m s^{-1}, how fast is the Earth moving towards it? FIGURE 2 shows the situation. The mass of the Earth is 6.0×10^{24} kg. We have:

total momentum of Earth and rock = 0

Hence:

$(60 \times 20) + (6.0 \times 10^{24} \times v) = 0 \Rightarrow v = -2.0 \times 10^{-22}$ m/s

The minus sign shows that the Earth's velocity is in the opposite direction to that of the rock. The Earth moves very slowly indeed. In the time of the rock's fall, it will move much less than the diameter of the nucleus of an atom!

EXAMPLE 72. RECOIL SPEED OF A RIFLE A 30 g ball is fired from a 1.2 kg spring-loaded toy rifle with a speed of 15 m/s. What is the recoil speed of the rifle?

APPROACH As the ball moves down the barrel, there are complicated forces exerted on the ball and on the rifle. However, if we take the system to be the ball + rifle, these are *internal* forces that do not change the total momentum.

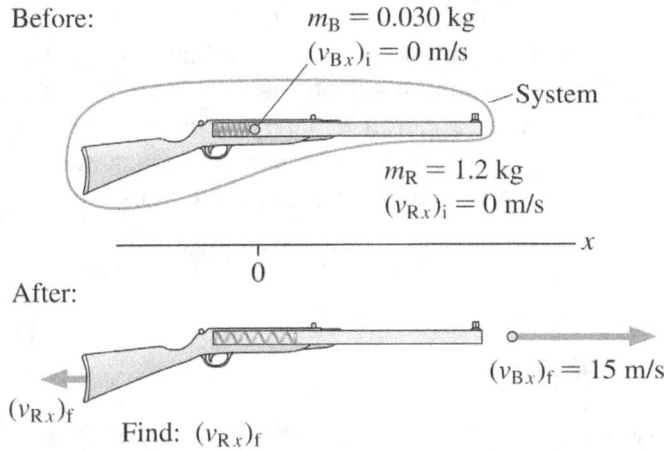

FIGURE 1. Before-and-after visual overview for a toy rifle.

The *external* forces of the rifle's and ball's weights are balanced by the external force exerted by the person holding the rifle, so $\sum \vec{F}_{ext} = 0$. This is an isolated system and the law of conservation of momentum applies.
FIGURE 1 shows a visual overview before and after the ball is fired. We'll assume the ball is fired in the +x-direction.
SOLUTION The x-component of the total momentum is $P_x = p_{Bx} + p_{Rx}$. Everything is at rest before the trigger is pulled, so the initial momentum is zero. After the trigger is pulled, the internal force of the spring pushes the ball down the barrel *and* pushes the rifle backward. Conservation of momentum gives
$(p_x)_f = m_B(v_{Bx})_f + m_R(v_{Rx})_f = (P_x)_i = 0$
Solving for the rifle's velocity, we find
$(v_{Rx})_f = -\frac{m_B}{m_R}(v_{Bx})_f = -\frac{0.030\,kg}{1.2\,kg} \times 15\ m/s = -0.38\,m/s$
The minus sign indicates that the rifle's recoil is to the left. The recoil *speed* is 0.38 m/s.
DISCUSSION Real rifles fire their bullets at much higher velocities, and their recoil is correspondingly higher. Shooters need to brace themselves against the "kick" of the rifle back against their shoulder.
EXAMPLE 73. A block of mass *m* is at rest in a gravity free space suddenly explodes into two parts in the ratio 1 : 3, if lighter particle has a speed of 9 m/s just after explosion, then find the speed of heavier particle.
SOLUTION $\vec{p}_{initial} = 0$, $\vec{p}_{final} = \left(\frac{m}{4}\right)9\hat{\imath} + \left(\frac{3m}{4}\right)\vec{v}$
By conservation of linear momentum, we have
$\vec{p}_{initial} = \vec{p}_{final}$
$0 = \frac{m}{4}9\hat{\imath} + \frac{3m}{4}\vec{v}$ or $\vec{v} = -3\hat{\imath}$ m/s

37. CHECKPOINT 8

1. ••A tennis ball of mass 57.0 g is held just above a basketball of mass 590 g. With their centers vertically aligned, both balls are released from rest at the same time, to fall through a distance of 1.20 m, as shown in FIGURE P1. (a) Find the magnitude of the downward velocity with which the basketball reaches the ground. (b) Assume that an elastic collision with the ground instantaneously reverses the velocity of the basketball while the tennis ball is still moving down. Next, the two balls meet in an elastic collision. To what height does the tennis ball rebound?

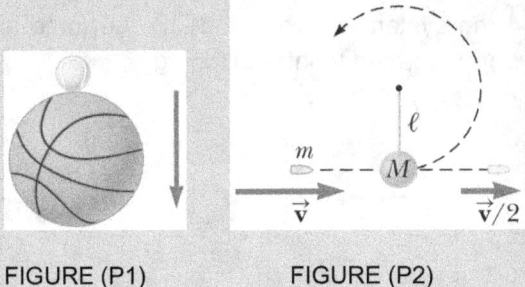

FIGURE (P1)   FIGURE (P2)

2. ••As shown in FIGURE P2, a bullet of mass m and speed v passes completely through a pendulum bob of mass M. The bullet emerges with a speed of $v/2$. The pendulum bob is suspended by a stiff rod (not a string) of length l, and negligible mass. What is the minimum value of v such that the pendulum bob will barely swing through a complete vertical circle?

3. ••Two blocks are free to slide along the frictionless, wooden track shown in FIGURE P3. The block of mass $m_1 = 5.00$ kg is released from the position shown, at height $h = 5.00$ m above the flat part of the track. Protruding from its front end is the north pole of a strong magnet, which repels the north pole of an identical magnet embedded in the back end of the block of mass $m_2 = 10.0$ kg, initially at rest. The two blocks never touch. Calculate the maximum height to which m_1 rises after the elastic collision.

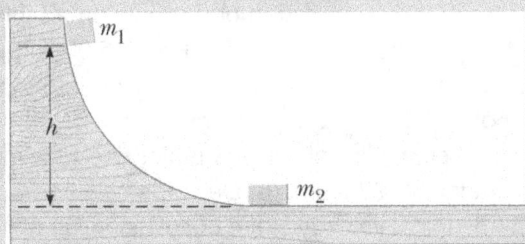

FIGURE (P3)

4. •••FIGURE P4a shows an overhead view of the initial configuration of two pucks of mass *m* on

frictionless ice. The pucks are tied together with a string of length l, and negligible mass. At time $t = 0$, a constant force of magnitude F begins to pull to the right on the center point of the string. At time t, the moving pucks strike each other and stick together. At this time, the force has moved through a distance d, and the pucks have attained a speed v (Fig. P4b). (a) What is v in terms of $F, d, l,$ and m? (b) How much of the energy transferred into the system by work done by the force has been transformed to internal energy?

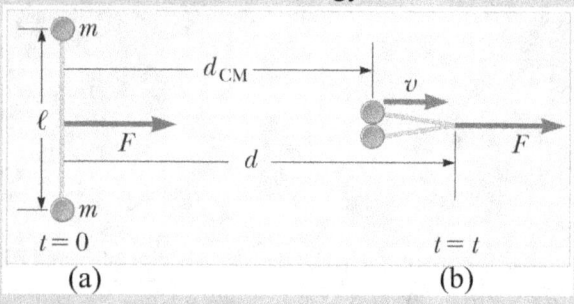

FIGURE (P4)

5. ••A wooden block of mass M rests on a table over a large hole as in FIGURE P5. A bullet of mass m with an initial velocity of v_i is fired upward into the bottom of the block and remains in the block after the collision. The block and bullet rise to a maximum height of h. (a) Describe how you would find the initial velocity of the bullet using ideas you have learned in this chapter. (b) Find an expression for the initial velocity of the bullet.

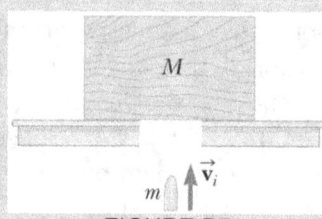

FIGURE P5

6. ••A 1.25-kg wooden block rests on a table over a large hole as in FIGURE P5. A 5.00-g bullet with an initial velocity vi is fired upward into the bottom of the block and remains in the block after the collision. The block and bullet rise to a maximum height of 22.0 cm. (a) Describe how you would find the initial velocity of the bullet using ideas you have learned in this chapter. (b) Calculate the initial velocity of the bullet from the information provided.

7. •••Two blocks of masses $m_1 = 2.00$ kg and $m_2 =$ 4.00 kg, are released from rest at a height of $h = 5.00$ m on a frictionless track as shown in FIGURE P6. When they meet on the level portion of the track, they undergo a head-on, elastic collision. Determine the maximum heights to which m_1 and m_2 rise on the curved portion of the track after the collision.

FIGURE P7

8. ••A bullet of mass $m = 8.00$ g is fired into a block of mass $M = 250$ g that is initially at rest at the edge of a table of height $h = 1.00$ m (Fig. P8). The bullet remains in the block, and after the impact the block lands $d = 2.00$ m from the bottom of the table. Determine the initial speed of the bullet.

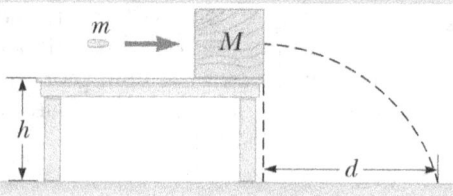

FIGURE P8

9. •••In FIGURE. P9, block 1 of mass m_1 slides from rest along a frictionless ramp from height $h = 2.50$ m and then collides with stationary block 2, which has mass $m_2 = 2.00m_1$. After the collision, block 2 slides into a region where the coefficient of kinetic friction $\mu_k = 0.500$ and comes to a stop in distance d within that region. What is the value of distance d if the collision is (a) elastic and (b) completely inelastic?

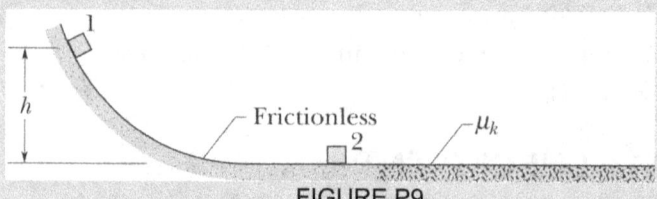

FIGURE P9

10. •••A cannon is rigidly attached to a carriage, which can move along horizontal rails but is connected to a post by a large spring, initially

unstretched and with force constant $k = 2.00 \times 10^4$ N/m, as shown in FIGURE P10. The cannon fires a 200-kg projectile at a velocity of 125 m/s directed 45.0° above the horizontal. (a) Assuming that the mass of the cannon and its carriage is 5 000 kg, find the recoil speed of the cannon. (b) Determine the maximum extension of the spring. (c) Find the maximum force the spring exerts on the carriage. (d) Consider the system consisting of the cannon, carriage, and projectile. Is the momentum of this system conserved during the firing? Why or why not?

FIGURE P10

11. ●●Show that if a body A collides elastically with another body of same mass at rest obliquely, then after the collision the two bodies move at right angles to each other, i. e. $\theta + \phi = \frac{\pi}{2}$.

38. QUESTIONS AND EXERCISES

38.1. CONCEPTUAL QUESTIONS (CQs)

1. ●●You are trapped on the second floor of a burning building. The stairway is impassable, but there is a balcony outside your window. Describe what might happen in the following situations. (a) You jump from the second story balcony to the pavement below, landing stiff legged on your feet. (b) You jump into a privet hedge, landing on your back and rolling to your feet. (c) You jump into a firefighters' net, landing on your back. What happens to the net as you land in it? What do the firefighters do to cushion your fall even more?

2. ●●An astronaut in deep space is taking a spacewalk when the tether connecting him to his spaceship breaks. How can he get back to the ship? He doesn't have a rocket propulsion backpack, unfortunately, but he is carrying a big wrench.

3. ●Which would be more effective: a hammer that collides *elastically* with a nail, or one that collides perfectly *inelastically*? Assume that the mass of the hammer is much larger than that of the nail.

4. ●In your own words, phrase each of Newton's three laws of motion as a statement about momentum.

5. ●The momentum of a system can only be changed by an external force. What is the external force that changes the momentum of a bicycle (with its rider) as it speeds up, slows down, or changes direction? Is it true that changes in the bicycle's kinetic energy must come from an external force? Explain.

6. ●Mohan and Raju are new to the sport of rock climbing. Mohan says he wants a stiff rope because a stiff rope is a strong rope. Raju insists that a good climbing rope must have some stretch. Who is correct, and why?

7. ●If you drop your keys, their momentum increases as they fall. Why is the momentum of the keys not conserved? Does this mean that the momentum of the universe increases as the keys fall? Explain.

8. ●A system of particles is known to have zero momentum. Does it follow that the kinetic energy of the system is also zero? Explain.

9. ●On a calm day you connect an electric fan to a battery on your sailboat and generate a breeze. Can the wind produced by the fan be used to power the sailboat? Explain.

10. ●Crash statistics show that it is safer to be riding in a heavy car in an accident than in a light car. Explain in terms of physical principles.

11. ● (a) As you approach a stoplight, you apply the brakes and bring your car to rest. What happened to your car's initial momentum? (b) When the light turns green, you accelerate until you reach cruising speed. What force was responsible for increasing your car's momentum?

12. ●An object at rest on a frictionless surface is struck by a second object. Is it possible for both objects to be at rest after the collision? Explain.

13. ●Two cars collide at an intersection. If the cars do not stick together, can we conclude that their collision was elastic? Explain.

14. ●An hourglass is turned over, and the sand is allowed to pour from the upper half of the glass to the lower half. If the hourglass is resting on a scale, and the total mass of the hourglass and sand is M, describe the reading on the scale as the sand

runs to the bottom.

15. •A tall, slender drinking glass with a thin base is initially empty. (a) Where is the center of mass of the glass? (b) Suppose the glass is now filled slowly with water until it is completely full. Describe the position and motion of the center of mass during the filling process.
16. •Lifting one foot into the air, you balance on the other foot. What can you say about the location of your center of mass?
17. •In the "Fosbury flop" method of high jumping, named for the track and field star Dick Fosbury, an athlete's center of mass may pass under the bar while the athlete's body passes over the bar. Explain how this is possible?

38.2. PRACTICE PROBLEMS (PPPs)

1. ••What impulse does the force shown in FIGURE PP1 exert on a 250 g particle?

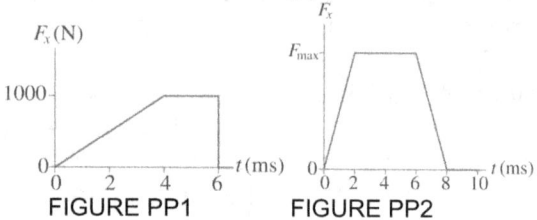

FIGURE PP1   FIGURE PP2

2. ••In FIGURE PP2, what value of F_{max} gives an impulse of 6.0 N s?
3. ••FIGURE PP3 is an incomplete momentum bar chart for a collision that lasts 10 ms. What are the magnitude and direction of the average collision force exerted on the object?

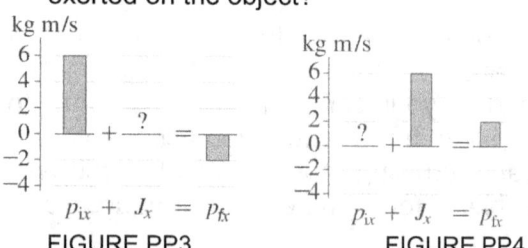

FIGURE PP3   FIGURE PP4

4. ••FIGURE PP4 is an incomplete momentum bar chart for a 50 g particle that experiences an impulse lasting 10 ms. What were the speed and direction of the particle before the impulse?
5. ••Far in space, where gravity is negligible, a 425 kg rocket traveling at 75 m/s fires its engines. FIGURE PP5 shows the thrust force as a function of time. The mass lost by the rocket during these 30 s is negligible. (a) What impulse does the engine impart to the rocket? (b) At what time does the rocket reach its maximum speed? What is the maximum speed?

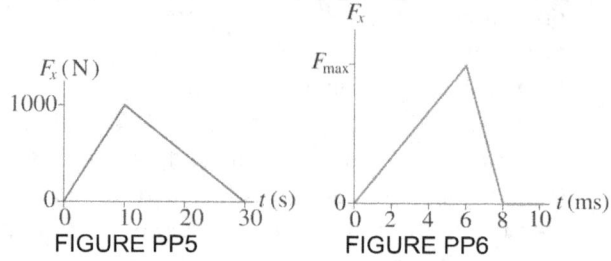

FIGURE PP5   FIGURE PP6

6. ••In FIGURE PP6, what value of F_{max} gives an impulse of 6.0 N.s?
7. ••A 600 g air-track glider collides with a spring at one end of the track. FIGURE PP7 shows the glider's velocity and the force exerted on the glider by the spring. How long is the glider in contact with the spring?

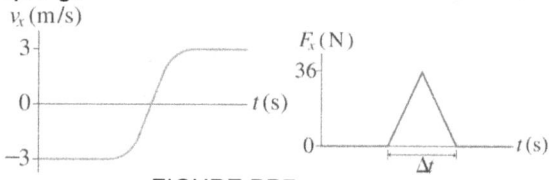

FIGURE PP7

8. ••A 250 g ball collides with a wall. FIGURE PP8 shows the ball's velocity and the force exerted on the ball by the wall. What is v_{fx}, the ball's rebound velocity?

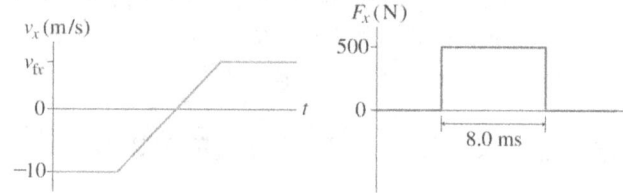

FIGURE PP8

9. ••A 12-kg hammer strikes a nail at a velocity of 8.5 m/s and comes to rest in a time interval of 8.0 ms. (a) What is the impulse given to the nail? (b) What is the average force acting on the nail?
10. ••A tennis ball of mass $m = 0.060$ kg and speed $v = 25$ m/s strikes a wall at a 45° angle and rebounds with the same speed at 45° (FIGURE PP10). What is the impulse (magnitude and direction) given to the ball?
11. ••A 5000 kg open train car is rolling on frictionless rails at 22 m/s when it starts pouring rain. A few minutes later, the car's speed is 20 m/s. What mass of water has collected in the car?

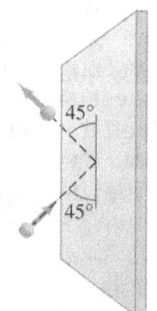

FIGURE PP10

12. ••A 12,000 kg railroad car is rolling at 3 m/s when a 3000 kg load of gravel is suddenly dropped in. What is the car's speed just after the gravel is loaded?
13. ••Three identical train cars, coupled together, are rolling east at speed v_0. A fourth car traveling east at

$2v_0$ catches up with the three and couples to make a four-car train. A moment later, the train cars hit a fifth car that was at rest on the tracks, and it couples to make a five-car train. What is the speed of the five-car train?

14. ●●A child in a boat throws a 5.70-kg package out horizontally with a speed of 10 m/s. Calculate the velocity of the boat immediately after, assuming it was initially at rest. The mass of the child is 24.0 kg and that of the boat is 35.0 kg.

15. ●●A 22-g bullet traveling penetrates a 2.0-kg block of wood and emerges going If the block is stationary on a frictionless surface when hit, how fast does it move after the bullet emerges?

16. ●●A mass $m_A = 2.0$ kg moving with velocity $\vec{v}_A = (4.0\hat{i} + 5.0\hat{j} - 2.0\hat{k})$ collides with mass $m_B = 3.0$ kg, which is initially at rest. Immediately after the collision, mass m_A is observed traveling at velocity $\vec{v}_A = (-2.0\hat{i} + 3.0\hat{k})$ m/s Find the velocity of mass after the collision. Assume no outside force acts on the two masses during the collision.

17. ●●A 224-kg projectile, fired with a speed of at a 60.0° angle, breaks into three pieces of equal mass at the highest point of its arc (where its velocity is horizontal). Two of the fragments move with the same speed right after the explosion as the entire projectile had just before the explosion; one of these moves vertically downward and the other horizontally. Determine (a) the velocity of the third fragment immediately after the explosion and (b) the energy released in the explosion.

18. ●●A 3.0-kg block slides along a frictionless tabletop at 8.0 m/s toward a second block (at rest) of mass 4.5 kg. A coil spring, which obeys Hooke's law and has spring constant $k = 850$ N/m is attached to the second block in such a way that it will be compressed when struck by the moving block, FIGURE PP18. (a) What will be the maximum compression of the spring? (b) What will be the final velocities of the blocks after the collision? (c) Is the collision elastic? Ignore the mass of the spring.

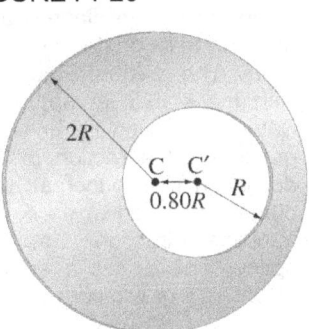

FIGURE PP18

19. ●●A 920-kg sports car collides into the rear end of a 2300-kg SUV stopped at a red light. The bumpers lock, the brakes are locked, and the two cars skid forward 2.8 m before stopping. The police officer, estimating the coefficient of kinetic friction between tires and road to be 0.80, calculates the speed of the sports car at impact. What was that speed?

20. ●●Two billiard balls of equal mass move at right angles and meet at the origin of an xy coordinate system. Initially ball A is moving along the y axis at +2 m/s and ball B is moving to the right along the x axis with speed +3.7 m/s. After the collision (assumed elastic), the second ball is moving along the positive y axis (FIGURE PP20). What is the final direction of ball A, and what are the speeds of the two balls?

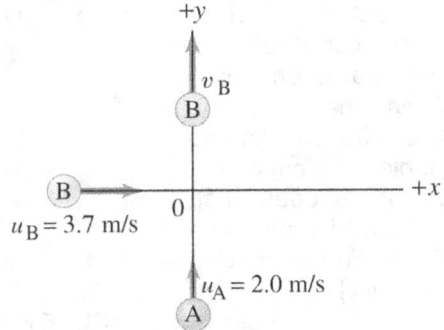

FIGURE PP20

21. ●●A uniform circular plate of radius 2R has a circular hole of radius R cut out of it. The center of the smaller circle is a distance 0.80R from the center C of the larger circle, FIGURE PP21. What is the position of the center of mass of the plate?

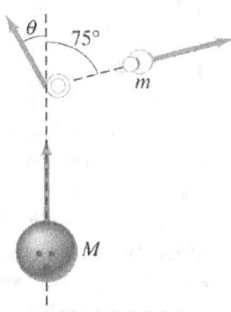

FIGURE PP21

22. ●●A 55-kg woman and a 72-kg man stand 10.0 m apart on frictionless ice. (a) How far from the woman is their CM? (b) If each holds one end of a rope, and the man pulls on the rope so that he moves 2.5 m, how far from the woman will he be now? (c) How far will the man have moved when he collides with the woman?

23. ●●●In order to convert a tough split in bowling, it is necessary to strike the pin a glancing blow as shown in FIGURE PP23. Assume that the bowling ball, initially traveling at 13 m/s has five times the mass of a pin and that the pin goes off at 75° from the original direction of the ball. Calculate the speed (a) of the pin and (b) of the ball just after collision, and (c) calculate the angle through which the ball was deflected. Assume the collision is elastic and ignore any spin of the ball.

FIGURE PP23

24. ●●●A 4800-kg open railroad car coasts along with a constant speed of on a level track. Snow begins to fall vertically and fills the car at a rate of 3.80 kg/min.

Ignoring friction with the tracks, what is the speed of the car after 60.0 min?

25. ●●In a physics lab, a cube slides down a frictionless incline as shown in FIGURE PP22 and elastically strikes another cube at the bottom that is only one-half its mass. If the incline is 35 cm high and the table is 95 cm off the floor, where does each cube land? [Both leave the incline moving horizontally]

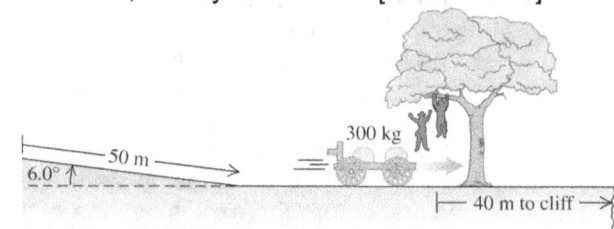

FIGURE PP25

26. ●●*The gravitational slingshot effect.* FIGURE PP26 shows the planet Saturn moving in the negative x direction at its orbital speed (with respect to the Sun) of 9.6 KM/S. The mass of Saturn is $5.69 \times 10^{26} kg$. A spacecraft with mass 825 kg approaches Saturn. When far from Saturn, it moves in the $+x$ direction at 10.4 km/s. The gravitational attraction of Saturn (a conservative force) acting on the spacecraft causes it to swing around the planet (orbit shown as dashed line) and head off in the opposite direction. Estimate the final speed of the spacecraft after it is far enough away to be considered free of Saturn's gravitational pull.

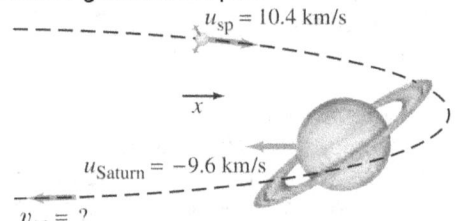

FIGURE PP26

27. ●●A 20.00-kg lead sphere is hanging from a hook by a thin wire 3.50 m long and is free to swing in a complete circle. Suddenly it is struck horizontally by a 5.00-kg steel dart that embeds itself in the lead sphere. What must be the minimum initial speed of the dart so that the combination makes a complete circular loop after the collision?

28. ●●An 8.00-kg ball, hanging from the ceiling by a light wire 135 cm long, is struck in an elastic collision by a 2.00-kg ball moving horizontally at 5.00 m/s just before the collision. Find the tension in the wire just after the collision.

29. ●●A fireworks rocket is fired vertically upward. At its maximum height of 80.0 m, it explodes and breaks into two pieces: one with mass 1.40 kg and the other with mass 0.28 kg. In the explosion, 860 J of chemical energy is converted to kinetic energy of the two fragments. (a) What is the speed of each fragment just after the explosion? (b) It is observed that the two fragments hit the ground at the same time. What is the distance between the points on the ground where they land? Assume that the ground is level and air resistance can be ignored.

30. ●●●A wagon with two boxes of gold, having total mass 300 kg, is cut loose from the horses by an outlaw when the wagon is at rest 50 m up a 6° slope (FIGURE PP30). The outlaw plans to have the wagon roll down the slope and across the level ground, and then fall into a canyon where his confederates wait. But in a tree 40 m from the canyon edge wait the Rahul (mass 75.0 kg) and Kapil (mass 60.0 kg). They drop vertically into the wagon as it passes beneath them. (a) If they require 5.0 s to grab the gold and jump out, will they make it before the wagon goes over the edge? The wagon rolls with negligible friction. (b) When the two heroes drop into the wagon, is the kinetic energy of the system of the heroes plus the wagon conserved? If not, does it increase or decrease, and by how much? [sin 6° = 0.10]

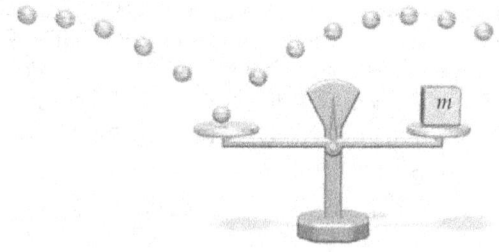

FIGURE PP30

31. ●●●A stream of elastic glass beads, each with a mass of 0.50 g, comes out of a horizontal tube at a rate of 100 per second (see FIGURE PP31). The beads fall a distance of $0.50\ m$ to a balance pan and bounce back to their original height. How much mass must be placed in the other pan of the balance to keep the pointer at zero?

FIGURE PP31

32. ●●A dumbbell consisting of two balls of mass m connected by a massless 1.00-m-long rod rests on a frictionless floor against a frictionless wall with one ball directly above the other. The center to-center distance between the balls is equal to 1.00 m. The dumbbell then

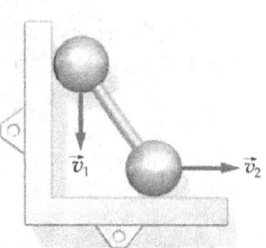

FIGURE PP32

begins to slide down the wall, as in FIGURE PP32. Find the speed of the bottom ball at the moment when it equals the speed of the top ball.

33. •••A cylindrical storage tank is initially filled with aviation gasoline. The tank is then drained through a valve on the bottom. See FIGURE PP33. (a) As the gasoline is withdrawn, describe qualitatively the motion of the center of mass of the tank and its remaining contents. (b) What is the depth x to which the tank is filled when the center of mass of the tank and its remaining contents reaches its lowest point? Express your answer in terms of H, the height of the tank; M, its mass; and m, the mass of gasoline it can hold.

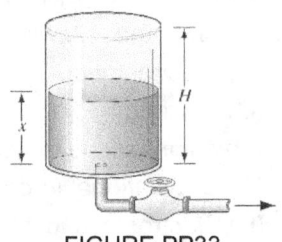

FIGURE PP33

34. •••"Relative" is an important word. In FIGURE PP34, block L of mass $m_L = 1.00$ kg and block R of mass $m_R = 0.500$ kg are held in place with a compressed spring between them. When the blocks are released, the spring sends them sliding across a frictionless floor. (The spring has negligible mass and falls to the floor after the blocks leave it.) (a) If the spring gives block L a release speed of 1.20 m/s relative to the floor, how far does block R travel in the next 0.800 s? (b) If, instead, the spring gives block L a release speed of 1.20 m/s relative to the velocity that the spring gives block R, how far does block R travel in the next 0.800 s?

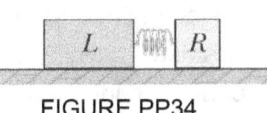

FIGURE PP34

35. •••In FIGURE PP35, block 1 of mass $m_1 = 6.6$ kg is at rest on a long frictionless table that is up against a wall. Block 2 of mass m_2 is placed between block 1 and the wall and sent sliding to the left, toward block 1, with constant speed u_2. Find the value of m_2 for which both blocks move with the same velocity after block 2 has collided once with block 1 and once with the wall. Assume all collisions are elastic (the collision with the wall does not change the speed of block 2).

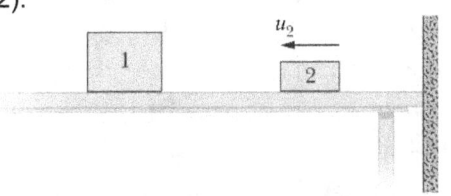

FIGURE PP35

36. •••The script for an action movie calls for a small race car (of mass 1500 kg and length 3.0 m) to accelerate along a flattop boat (of mass 4000 kg and length 14 m), from one end of the boat to the other, where the car will then jump the gap between the boat and a somewhat lower dock. You are the technical advisor for the movie. The boat will initially touch the dock, as in FIGURE PP36; the boat can slide through the water without significant resistance; both the car and the boat can be approximated as uniform in their mass distribution. Determine what the width of the gap will be just as the car is about to make the jump.

37. ••A fireworks rocket is moving at a speed of 45.0 m/s. The rocket suddenly breaks into two pieces of equal mass, which fly off with velocities $\vec{v}_1$ and $\vec{v}_2$, as shown in the drawing. What are the magnitudes of (a) $\vec{v}_1$ and (b) $\vec{v}_2$?

38. A particle at rest is constrained to move on a smooth horizontal surface. Another identical particle hits this stationary particle with a velocity v at an angle $\theta = 60°$ with horizontal. If the particles move together, then find the velocity of the combination just after the impact.

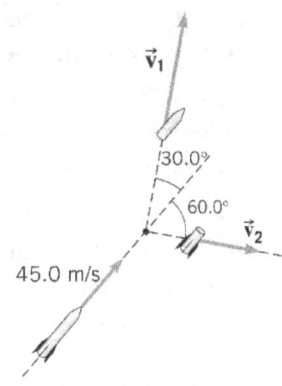

FIGURE PP36

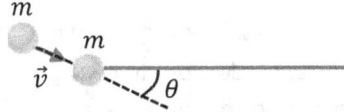

39. •• A fast-moving neutron suffers one-dimensional elastic collision with a nucleus $^{14}_{7}N$. What approximate percentage of energy is lost by the neutron in the collision?

40. •• A neutron mass $1.67 \times 10^{-27} kg$ moving with a velocity $1.2 \times 10^7 ms^{-1}$ collides head-on with a deuteron of mass $3.34 \times 10^{-27} kg$ initially at rest. If the collision is perfectly inelastic, what will be the speed of the composite particle?

41. •• In above question, if the collision were perfectly elastic, what would be the speed of deuteron after the collision?

38.3. MULTIPLE-CHOICE ASSIGNMENTS

38.3.1. LEVEL 1

1.1.1.1. CONSERVATION OF LINEAR MOMENTUM

1. A bullet is fired from the gun. The gun recoils, the kinetic energy of the recoil shall be-
   (A) equal to the kinetic energy of the bullet

(B) less than the kinetic energy of the bullet
(C) greater than the kinetic energy of the bullet
(D) double that of the kinetic energy of the bullet

2. A bomb at rest explodes into two parts of masses m_1 and m_2. If the momentums of the two parts be p_1 and p_2, then their kinetic energies will be in the ratio of-
(A) m_1/m_2 (B) m_2/m_1
(C) p_1/p_2 (D) p_2/p_1

3. Conservation of linear momentum is equivalent to-
(A) Newton's second law of motion
(B) Newton's first law of motion
(C) Newton's third law of motion
(D) Conservation of angular momentum.

4. A body of mass m collides against a wall with the velocity v and rebounds with the same speed. Its change of momentum is-
(A) $2mv$ (B) mv (C) $-mv$ (D) 0

5. A bomb initially at rest explodes by itself into three equal mass fragments. The velocities of two fragments are $(3\hat{i} + 2\hat{j})$ m/s and $(-\hat{i} - 4\hat{j})$ m/s. The velocity of the third fragment is (in m/s)-
(A) $2\hat{i} + 2\hat{j}$ (B) $2\hat{i} - 2\hat{j}$
(C) $-2\hat{i} + 2\hat{j}$ (D) $-2\hat{i} - 2\hat{j}$

6. A stone of mass m_1 moving with a uniform speed v suddenly explodes on its own into two fragments. If the fragment of mass m_2 is at rest, the speed of the other fragment is-
(A) $\frac{m_1 v}{(m_1-m_2)}$ (B) $\frac{m_2 v}{(m_1-m_2)}$
(C) $\frac{m_1 v}{(m_1+m_2)}$ (D) $\frac{m_1 v}{m_2}$

7. A monkey of mass 20kg rides on a 40kg trolley moving with constant speed of 8m/s along a horizontal track. If the monkey jumps vertically to grab the overhanging branch of a tree, the speed of the trolley after the monkey has jumped off is –
(A) 8 m/s   (B) 1 m/s
(C) 4 m/s   (D) 12 m/s

8. A nucleus of mass number A originally at rest emits α-particle with speed v. The recoil speed of daughter nucleus is:
(A) $\frac{4v}{A-4}$ (B) $\frac{4v}{A+4}$ (C) $\frac{v}{A-4}$ (D) $\frac{v}{A+4}$

9. Two particles A and B which are initially at rest move towards each other under the mutual force of attraction. At the instant when the speed of A is v and the speed of B is $2v$. The speed of the centre of mass of the system is –
(A) v (B) $1.5v$ (C) $3v$ (D) zero

10. Which of the following force is conservation force –
(A) Electrostatic   (B) Frictional
(C) Viscous   (D) Air resistance

11. Which one of the following force is non-conservative?
(A) Gravitational force
(B) Electrostatic force
(C) Lorentz force
(D) Viscous force

12. A body is dropped from a certain height. When it lost an amount of P.E. 'U', it acquires a velocity 'v'. The mass of the body is-
(A) $\frac{2U}{v^2}$ (B) $\frac{2v}{U^2}$ (C) $\frac{2v}{U}$ (D) $\frac{U^2}{2v}$

13. A lead bullet of specific heat 's' moving with a velocity v strikes a wall and stops. If half of its energy is converted into heat, the rise in its temperature will be- [where s is in cal/kg - °C]
(A) $\frac{v^3 s}{J}$ (B) $\frac{2v^2}{Js}$ (C) $\frac{v^2}{4Js}$ (D) $\frac{v^2 s}{2J}$

14. A block of mass m slides down along the surface of the bowl (radius R) from the rim to the bottom. The velocity of the block at the bottom will be-
(A) $\sqrt{(\pi R g)}$ (B) $2\sqrt{(\pi R g)}$
(C) $\sqrt{2Rg}$ (D) $\sqrt{gR}$

15. A sphere is suspended by a thread of length l. What minimum horizontal velocity is to be imparted to the sphere for it to reach the height of suspension?
(A) $\sqrt{8l}$ (B) gl (C) $\sqrt{2gl}$ (D) $\sqrt{l/g}$

16. A body of mass m kg is rotating in a vertical circle at the end of a string of length r meter. The difference in the $K.E.$ at the top and bottom of the circle is-
(A) mgr (B) $2mgr$ (C) $\frac{mg}{r}$ (D) $\frac{2mg}{r}$

17. A ball of mass $2\,kg$ is projected horizontally with a velocity 20m/s from a building of height 15m. The speed with which body hits the ground is-
(A) $20\,m/s$ (B) $10(7)^{1/2}\,m/s$
(C) $25\frac{m}{s}$ (D) $35\frac{m}{s}$

18. A man slides down a snow covered hill along a curved path and falls $20m$ below his initial position. The velocity in m/sec with which he finally strikes the ground is ($g = 10\,m/s^2$)

(A) 20 (B) 400 (C) 200 (D) 40

19. When a 20 g mass hangs attached to one end of a light spring of length 10cm, the spring stretches by 2 cm. The mass is pulled down until the total length of the spring is 14 cm. The elastic energy, m Joule stored in the spring is-
   (A) 4×10^{-2}
   (B) 4×10^{-3}
   (C) 8×10^{-2}
   (D) 8×10^{-3}

1.1.1.2. CENTRE OF MASS

20. The velocity of centre of mass in absence of external force is –
   (A) constant
   (B) zero
   (C) increases
   (D) decreases

21. The centre of mass of two particles lies
   (A) on the line perpendicular to the line joining the particles
   (B) on a point outside the line joining the particles
   (C) on the line joining the particles.
   (D) none of the above.

22. Two particles A and B which are initially at rest move towards each other under the mutual force of attraction. At the instant when the speed of A is v and the speed of B is $2v$. the speed of the centre of mass of the system is-
   (A) v (B) $1.5v$ (C) $3v$ (D) zero

23. Two particles whose masses are 10 kg and 30 kg and their position vectors are $\hat{i}+\hat{j}+\hat{k}$ and $-\hat{i}-\hat{j}-\hat{k}$ respectively would have the centre of mass at-
   (A) $\frac{(\hat{i}+\hat{j}+\hat{k})}{2}$
   (B) $\frac{(\hat{i}+\hat{j}+\hat{k})}{2}$
   (C) $\frac{(\hat{i}+\hat{j}+\hat{k})}{4}$
   (D) $\frac{(\hat{i}+\hat{j}+\hat{k})}{4}$

24. Two balls A and B of masses 100 gm and 250 gm respectively are connected by a stretched spring of negligible mass and placed on a smooth table. When the balls are released simultaneously the initial acceleration of B is 10 cm/s^2 west ward. What is the magnitude and direction of initial acceleration of the ball A-
   (A) 25 cm/sec^2 Eastward
   (B) 25 cm/sec^2 North ward
   (C) 25 cm/sec^2 West ward
   (D) 25 cm/sec^2 South ward

25. A uniform square plate ABCD has a mass of 10kg. If two point masses of 3 kg each are placed at the corners C and D as shown in the adjoining figure, then the centre of mass shifts to the point which is lie on –

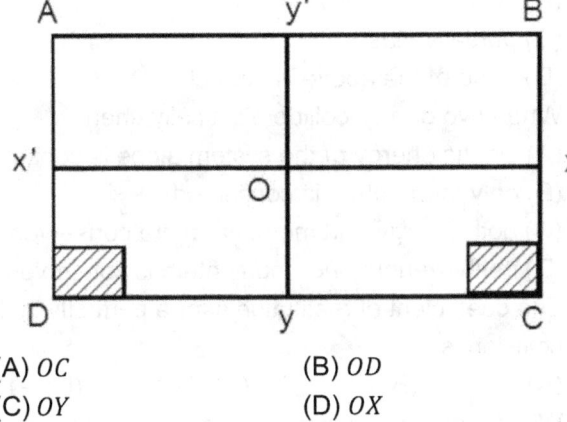

   (A) OC
   (B) OD
   (C) OY
   (D) OX

26. The velocity of centre of mass in absence of external force is –
   (A) constant
   (B) zero
   (C) increases
   (D) decreases

1.1.1.3. COLLISION

27. In an inelastic collision-
   (A) momentum is conserved but kinetic energy is not
   (B) momentum is not conserved but kinetic energy is conserved
   (C) neither momentum nor kinetic energy is conserved
   (D) both the momentum and kinetic energy are conserved

28. Inelastic collision is the-
   (A) collision of ideal gas molecule with the walls of the container
   (B) collision of electron and positron to an inhilate each other.
   (C) collision of two rigid solid spheres lying on a frictionless table
   (D) scattering of α-particles with the nucleus of gold atom

29. Which of the following is not a perfectly inelastic collision-
   (A) capture of an electron by proton
   (B) man jumping onto a moving cart
   (C) collision between glass balls
   (D) a bullet fired into a block of wood such that it is embedded in the wood

30. When two bodies stick together after collision, the collision is said to be
    (A) inelastic
    (B) elastic
    (C) partially elastic
    (D) none of the above is correct
31. When two bodies collide elastically, then
    (A) kinetic energy of the system alone is conserved
    (B) only momentum is conserved
    (C) both energy and momentum are conserved
    (D) neither energy nor momentum is conserved
32. The coefficient of restitution e for a perfectly inelastic collision is-
    (A) 1   (B) ∞   (C) Zero   (D) −1
33. Which of the following statements is true for collisions-
    (A) momentum is conserved in elastic collisions but not in inelastic collisions.
    (B) total-kinetic energy is conserved in elastic collisions but momentum is not conserved.
    (C) total kinetic energy is not conserved in inelastic collisions but momentum is conserved
    (D) total kinetic energy and momentum both are conserved in all types of collisions
34. A ball hits the floor and rebounds after an inelastic collision. In this case-
    (A) the momentum of the ball just after the collision is the same as that just before the collision
    (B) the mechanical energy of the ball remains the same in the collision
    (C) the total momentum of the ball and the earth is conserved.
    (D) the total energy of the ball and the earth is conserved
35. Consider the elastic collision of two bodies A and B of equal mass. Initially B is at rest and A moves with velocity v. After the collision-
    (A) the body A traces, its path back with the same speed
    (B) the body A comes to rest and B moves always in the direction of A's approach with the velocity v
    (C) both the bodies stick together and are at rest
    (D) B moves along with velocity $v/2$ and A retraces its path with velocity $v/2$
36. A particle A suffers an oblique elastic collision with a particle B that is at rest initially. If their masses are the same, then after the collision-
    (A) they will move in opposite directions
    (B) A continuous to move in the original direction while B remains at rest
    (C) they will move in mutually perpendicular direction
    (D) A comes to rest and B starts moving in the direction of the original motion of A
37. Five identical elastic balls are so suspended with strings of equal length in a row that the distances between adjacent balls are very small. If the extreme right ball is moved aside and released, then-

    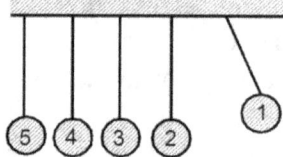

    (A) one extreme left hand ball will bounce off.
    (B) two extreme left hand balls will bounce off.
    (C) three extreme left hand balls will bounce off
    (D) all the left hand four balls will bounce off.
38. Six steel balls of identical size are lined up long a straight frictionless groove. Two similar balls moving with a speed V along the groove collide with this row on the extreme left hand then-
    (A) all the balls will start moving to the right with speed 1/8 each
    (B) all the six balls initially at rest will move on with speed V/6 each and two identical balls will come to rest
    (C) two balls from the extreme right end will move on with speed V each and the remaining balls will remain at rest
    (D) one ball from the right end will move on with speed 2V, the remaining balls will be at rest.

    Frictionless Groove
39. A body of mass 'm' moving with a constant velocity v hits another body of the same mass moving with the same velocity v but in the opposite direction and sticks to it. The velocity of the compound body after collision is-
    (A) v (B) $2v$ (C) $v/2$ (D) 0

40. A body of mass 2 kg moving with a velocity of 3m/sec towards left collides head-on with a body of mass 3 kg moving in opposite direction with a velocity 2m/sec. After collision the two bodies stick together and move with a common velocity which is-
(A) 5m/sec towards left
(B) 12 m/sec. towards right
(C) 12/5 m/sec. towards left
(D) zero

41. A 1.0 kg ball drops vertically into a floor from a height of 25 cm. It rebounds to a height of 4 cm. The coefficient of restitution for the collision is-
(A) 0.16   (B) 0.32   (C) 0.40   (D) 0.56

42. An inelastic ball is dropped from a height 100 metre. If due to impact it loses 35% of its energy the ball will rise to a height of –
(A) 35 m   (B) 65 m   (C) 100 m   (D) 135 m

43. The bob of a simple pendulum of length l dropped from a horizontal position strikes a block of the same mass, placed on a horizontal table (frictionless) as shown in the diagram, the block shall have kinetic energy-

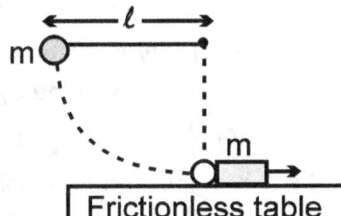

(A) Zero   (B) mgl (C) $1/2\, mgl$ (D) $2mgl$.

44. A particle of mass m_1 hits another particle of mass m_2 at rest with a velocity. The collision is head-on and elastic. If $m_1 \gg m_2$, then after collision, the velocity of m_2 will be-
(A) $\vec{u}$ (B) $-\vec{u}$ (C) $2\vec{u}$ (D) $-2\vec{u}$

45. A neutron travelling with a velocity v and K.E. E. collides perfectly elastically head on with the nucleus of an atom of mass number A at rest. The fraction of total energy retained by neutron is-
(A) $\left(\frac{A-1}{A+1}\right)^2$ (B) $\left(\frac{A+1}{A-1}\right)^2$ (C) $\left(\frac{A-1}{A}\right)^2$ (D) $\left(\frac{A+1}{A}\right)^2$

46. If one sphere collides head on with another sphere of the same mass at rest inelastically. The ratio of their speeds after collision shall be-
(A) $(1-e)/(1+e)$.   (B) $2e/(1+e)$
(C) $(1+e)/(1-e)$.   (D) e.

47. An object of mass 5 kg and speed 10 ms^{-1} explodes into two pieces of equal mass. One piece comes to rest. The kinetic energy added to the system during the explosion is-
(A) Zero.   (B) 50 J.   (C) 250 J.   (D) 500 J.

48. Two particles of same mass m moving with velocities u_1 and u_2 collide perfectly inelastically. The loss of energy would be-
(A) $\frac{1}{2}m(u_1-u_2)^2$ (B) $\frac{1}{4}m(u_1-u_2)^2$
(C) $m(u_1-u_2)^2$.   (D) $2m(u_2-u_1)^2$.

49. A particle of mass m_1 moving with velocity v collides head-on with a particle of mass m_2 initially at rest. The collision is completely inelastic. The fraction of the original kinetic energy that is converted into heat is-
(A) $m_1/(m_1+m_2)$.   (B) $m_2/(m_1+m_2)$.
(C) $m_1/(m_1-m_2)$.   (D) $m_2/(m_1-m_2)$.

50. A ball of mass m moving with a speed u undergoes a head-on elastic collision with a ball of mass nm initially at rest. The fraction of the incident energy transferred to the heavier ball is-
(A) $\frac{n}{1+n}$ (B) $\frac{n}{(1+n)^2}$ (C) $\frac{2n}{(1+n)^2}$ (D) $\frac{4n}{(1+n)^2}$

51. Two elastic bodies P and Q having equal masses are moving along the same line with velocities of 16 m/s and 10 m/s respectively. Their velocities after the elastic collision will be in m/s –
(A) 0 and 25   (B) 5 and 20
(C) 10 and 16   (D) 20 and 5

52. Two solid balls of rubber A and B whose masses are 200 gm and 400 gm respectively, are moving in mutually opposite directions. If the velocity A is 0.3 m/s and both the balls come to rest after collision, then the velocity of ball B is –
(A) 0.15 ms^{-1} (B) $-0.15\, ms^{-1}$
(C) 1.5 ms^{-1} (D) none of these

53. Two similar balls P and Q having velocities of 0.5 m/s and -0.3 m/s respectively collide elastically. The velocities of P and Q after the collision will respectively be –
(A) -0.5 m/s and 0.3 m/s
(B) 0.5 m/s and 0.3 m/s
(C) -0.3 m/s and 0.5 m/s
(D) 0.3 m/s and 0.5 m/s

54. Which of the following does not hold when two particles of masses m_1 and m_2 undergo elastic collision?
(A) when $m_1 = m_2$ and m_2 is stationary, there is maximum transfer of kinetic energy in head on collision
(B) when $m_1 = m_2$ and m_2 is stationary, there is maximum transfer of momentum in head on collision
(C) when $m_1 \gg m_2$ and m_2 is stationary, after head on collision m_2 moves with twice the velocity of m_1.
(D) when the collision is oblique and $m_1 = m_2$ with m_2 stationary, after the collision the particle move in opposite directions.

55. As shown in figure A, B and C are identical balls B and C are at rest and, the ball A moving with velocity v collides elastically with ball B, then after collision:

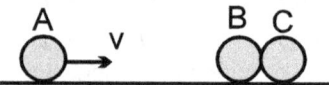

(A) All the three balls move with velocity $v/2$
(B) A comes to rest and $(B + C)$ moves with velocity $v/\sqrt{2}$
(C) A moves with velocity v and $(B + C)$ moves with velocity v
(D) A and B come to rest and C moves with velocity v

56. A moving sphere of mass m suffer a perfect elastic collision (not head on) with an equally massive stationary sphere. After collision both fly off at angle θ, value of which is:
(A) 0 (B) π
(C) indeterminate (D) $\pi/2$

57. A body of mass m kg collides elastically with another body at rest and then continues to move in the original direction with one half of its original speed. what is the mass of the target body?
(A) m kg (B) $2/3$ m kg (C) $m/3$ kg (D) $m/2$ kg

58. Two balls at the same temperature collide inelastically. What is conserved-
(A) momentum (B) velocity
(C) temperature (D) kinetic energy

59. A metal ball hits a wall and does not rebound whereas a rubber ball of the same mass on hitting the wall the same velocity rebounds back. It can be concluded that –

(A) metal ball suffers greater change in momentum
(B) rubber ball suffers greater change in momentum.
(C) the initial momentum of metal ball is greater than the initial momentum of rubber ball.
(D) both suffer same change in momentum.

60. Before a rubber ball bounces off from the floor the ball is in contact with the floor for a fraction of second. Which of the following statements are correct-?
(A) conservation of energy is not valid during this period
(B) conservation of energy is valid during this period
(C) as ball compressed kinetic energy is converted compressed potential energy
(D) 2 and 3 both

61. A ball of $0.1\ kg$ strikes a wall at right angle with a speed of 6 m/s and rebounds along its original path at 4 m/s. The change in momentum in Newton- sec is-
(A) 10^3 (B) 10^2 (C) 10 (D) 1

62. A particle of mass m moving with velocity v collides with particle of mass $2m$ at rest and adheres to it. The velocity of the system is-
(A) $2v$ (B) $3v$ (C) $v/2$ (D) $v/3$.

63. A bullet of mass a and velocity b is fired into a large block of mass c. The final velocity of the system is-
(A) $\frac{c}{a+b}b$ (B) $\frac{a}{a+c}b$ (C) $\frac{a+b}{c}\cdot a$ (D) $\frac{a+c}{a}\cdot b$

64. An elastic ball of mass m falls from a height h on an Aluminum disc of area A floating in a mercury pool. If the collision is perfectly elastic, the momentum transferred to the disc is-
(A) $\sqrt{2mgh}$ (B) $2\sqrt{mgh}$
(C) $m\sqrt{2gh}$ (D) $2m\sqrt{2gh}$

38.3.2. LEVEL 2

1. A particle moves in the x-y plane under the action of a force $\vec{F}$ such that the value of its linear momentum P at any time t is
$P_x = 2\cos t,\ P_y = 2\sin t$
the angle 'θ' between and at any given time t will be –
(A) 90° (B) 0°
(C) 180° (D) 30°

2. A spring is compressed between two toy-carts of masses m_1 and m_2. When the toy-carts are released

the spring exerts on each toy-cart equal and opposite force for the same time t. If the coefficient of friction 'μ' between the ground and the toy-carts are equal, then the displacement of the two toy carts

(A) $\frac{S_1}{S_2} = +\frac{m_2}{m_1}$ (B) $\frac{S_1}{S_2} = +\frac{m_1}{m_2}$
(C) $\frac{S_1}{S_2} = +\left(\frac{m_2}{m_1}\right)^2$ (D) $\frac{S_1}{S_2} = +\left(\frac{m_1}{m_2}\right)^2$

3. Sand drops fall vertically at the rate of $2\ kg/s$ on to a conveyor belt moving horizontally with the velocity of $0.2\ m/s$. Then the extra force needed to keep the belt moving is
(A) 0.4 newton (B) 0.08 newton
(C) 0.04 newton (D) 0.2 newton

4. An engine pumps a liquid of density 'ρ' continuously through a pipe of area of cross section A. If the speed with which the liquid passes through a pipe is v. then the rate at which the Kinetic energy is being imparted to the liquid is
(A) $A\rho v^3/2$ (B) $(1/2)A\rho v$
(C) $A\rho v^2/2$ (D) $A\rho v^2$

5. A boy is standing at the centre of a boat which is free to move on water. If the masses of the boy and the boat are m_1 and m_2 respectively and the boy moves a distance of $1\ m$ forward then the movement of the boat, in meter, is-
(A) $\frac{m_1}{m_1+m_2}$ (B) $\frac{m_2}{m_1+m_2}$
(C) $\frac{m_1}{m_2}$ (D) $\frac{m_2}{m_1}$

6. A bullet of mass m moving with velocity v_1 strikes a suspended wooden block of mass M as shown in the figure and sticks to it. If the block rises to a height y, the initial velocity of the bullet is –

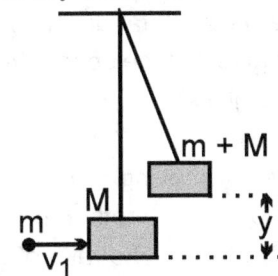

(A) $v_1 = \frac{m+M}{m}\sqrt{2gy}$ (B) $v_1 = \sqrt{2gy}$
(C) $v_1 = \frac{M+m}{M}\sqrt{2gy}$ (D) $v_1 = \frac{m}{m+M}\sqrt{2gy}$

7. A bullet of mass m strikes a pendulum bob of mass M with velocity u. It passes through and emerges out with a velocity u/2 from bob. The length of the pendulum is l. What should be the minimum value of u if the pendulum bob will swing through a complete circle?
(A) $\frac{2M}{m}\sqrt{5gl}$ (B) $\frac{M}{2m}\sqrt{5gl}$
(C) $\frac{2M}{m}\frac{1}{\sqrt{5gl}}$ (D) $\frac{M}{2m}\frac{1}{\sqrt{5gl}}$

8. An open water tight railway wagon of mass 5×10^3 kg coasts at an initial velocity of $1.2\ m/s$ without friction on a railway track. Rain falls vertically downwards into the wagon. What change then occurred in the kinetic energy of the wagon, when it has collected 10^3kg of water?
(A) 1200J (B) 300J (C) 600J (D) 900J

9. A ball moving on a horizontal frictionless plane hits an identical ball at rest with a velocity of 50cm/sec. If the collision is elastic, calculate the speed imparted to the target ball if the speed of the striking ball after the collision is 30cm/sec.
(A) 20 cm/sec (B) 30 cm/sec
(C) 40 cm/sec (D) 50 cm/sec

10. A ball is dropped from a height h on the ground. If the coefficient of restitution is e, the height to which the ball goes up after it rebounds for the nth time is.
(A) he^{2n} (B) $h\,e^n$ (C) $\frac{e^{2n}}{h}$ (D) $\frac{h}{e^{2n}}$

11. A ball after falling a distance of 5 metre from rest hits elastically the floor of a lift and rebounds. At the time of impact the lift was moving up with a velocity of 1m/sec. The velocity with which the ball rebounds just after impact is ($g = 10\ m/s^2$)
(A) $10\ m/sec$. (B) $11 m/sec$.
(C) $12\ m/sec$ (D) $13\ m/sec$.

12. Two equal lumps of putty are suspended side by side from two long strings so that they are just touching. One is drawn aside so that its centre of gravity rises a vertical distance h. It is released and then collides inelastically with the other one. The vertical distance risen by the centre of gravity of the combination is –
(A) h (B) $3h/4$ (C) $h/2$ (D) $h/4$

13. A particle of mass m collides perfectly elastically with another particle of mass M = 2m. If the incident particle is deflected by 90°. The heavy mass will make an angle with the initial direction of m equal to –
(A) 15° (B) 30º (C) 45º (D) 60º

14. A ball collides elastically with another ball of the same mass. The collision is oblique and initially one of the ball was at rest. After the collision, the two balls move with same speeds. What will be the angle between the velocity of the balls after the collision?
    (A) 30° (B) 45° (C) 60° (D) 90°

15. A billiard ball moving at a speed $2m/s$ strikes an identical ball initially at rest, at a glancing blow. After the collision one ball is found to be moving at a speed of $1m/s$ at 60° with the original line of motion. The velocity of the other ball shall be –
    (A) $(3)^{1/2} m/s$ at 30° to the original direction.
    (B) $1 m/s$ at 60° to the original direction.
    (B) $(3)^{1/2} m/s$ at 60° to the original direction.
    (D) $1 m/s$ at 30° to the original direction.

16. An explosion blows a rock into three paths. Two pieces go off at right angles. to each other. $1.00\ kg$ piece with a velocity $12\ m/sec$ and the other $2.00\ kg$ piece with a velocity $8\ m/sec$. If the third piece flies off with a velocity $40\ m/sec$. Then the mass of the third piece is –
    (A) $0.2\ kg$ (B) $0.3\ kg$ (C) $0.4\ kg$ (D) $0.5\ kg$

17. A proton of mass $1.67 \times 10^{-27} kg$ undergoes a head on collision with an α-particle initially at rest. After the collision, the α-particle moves with a speed of $8 \times 10^5\ m/sec$. Calculate the velocity of the proton before and after the collision. Given mass of α-particle $6.68 \times 10^{-27} kg$
    (A) $2 \times 10^3 m/sec, -1.2 \times 10^3 m/sec$
    (B) $2 \times 10^4 m/sec, -1.2 \times 10^4\ m/sec$
    (C) $2 \times 10^5 m/sec, -1.2 \times 10^5 m/sec$
    (D) $2 \times 10^6 m/sec, -1.2 \times 10^6\ m/sec$

18. A stationary body of mass m gets exploded in 3 parts having mass in the ratio of $1:3:3$. Its two fractions having equal mass moving at right angle to each other with velocity of $15\ m/sec$. Then the velocity of the third body is –
    (A) $45\sqrt{(2)}$ m/sec (B) $5\ m/sec$
    (C) $5\sqrt{(32)}$ m/sec (D) none of these

19. Three particles each of mass m are located at the vertices of an equilateral triangle ABC. They start moving with equal speeds v each along the medians of the triangle and collide at its centroid G. If after collision, A comes to rest and B retraces its path along GB, then C
    (A) also comes to rest
    (B) moves with a speed v along CG
    (C) moves with a speed v along BG
    (D) moves with a speed along AG

20. A cannon ball is fired with a velocity 200m/sec at an angle of 60° with the horizontal. At the highest point of its flight. It explodes into 3 equal fragments, one going vertically upwards with a velocity $100\ m/sec$, the second one falling vertically downwards with a velocity $100\ m/sec$. The third fragment will be moving with a velocity
    (A) $100\ m/sec$ in the horizontal direction
    (B) $300 m/sec$ in the horizontal direction
    (C) $300\ m/sec$ in a direction making an angle of 60° with the horizontal
    (D) $200\ m/sec$ in a direction making an angle of 60° with the horizontal

21. A shell lying in a smooth horizontal tube suddenly explodes and breaks of masses m_1 and m_2. If x is the distance of separation in the tube of the masses after time t seconds. Then energy released by explosion is –
    (A) $\frac{2m_1 m_2 x^2}{(m_1+m_2)t^2}$
    (B) $\frac{m_1 m_2 t^2}{2x^2(m_1+m_2)}$
    (C) $\frac{m_1 m_2 x^2}{2(m_1+m_2)t^2}$
    (D) None of these

22. A small ball of mass $m = 1\ gm$ is placed at the bottom of a spherical glass of radius $R = 1m$. It is displaced by height, $h = 1\ cm$ long the glass surface and released. What is the total distance described by it before coming to rest at the bottom ($\mu = 0.1$ between the wall and the glass)
    (A) $16\ cm$ (B) $7\ cm$ (C) $10\ cm$ (D) $8\ cm$

23. A body of mass m moving with velocity v makes a head-on collision with another body of mass $2m$ which is initially at rest. The ratio of kinetic energies of colliding body before and after collision will be-
    (A) $9:1$ (B) $1:1$ (C) $4:1$ (D) $2:1$

24. A lead ball of mass $2\ kg$ moving with a velocity of $1.5\ ms^{-1}$ hits against a ball of mass $3\ kg$ at rest. If the second ball moves with a velocity of $1 ms^{-1}$ after the impact in the original direction of motion of the first ball, the loss of K.E. due to impact is-

(A) 0.033 J (B) 0.75 J
(C) 1.5 J (D) 2.25 J

25. A billiard ball moving at a speed of $6.6\ ms^{-1}$ strikes an identical stationary ball a glancing blow. After the collision, one ball is found to be moving at a speed of $3.3\ ms^{-1}$ in a direction making an angle of 60º with the original line of motion. The velocity of the other ball is
    (A) $4.4\ ms^{-1}$ (B) $6.6\ ms^{-1}$
    (C) $3.3\ ms^{-1}$ (D) $5.7\ ms^{-1}$

38.3.3. LEVEL 3

1. A block of mass $m = 0.1\ kg$ is released from a height of 4 m on a curved smooth surface. On the horizontal surface, path AB is smooth and path BC offers coefficient of friction $\mu = 0.1$. If the impact of block with the vertical wall at C be perfectly elastic, the total distance covered by the block on the horizontal surface before coming to rest will be (take $g = 10\ ms^{-2}$)-

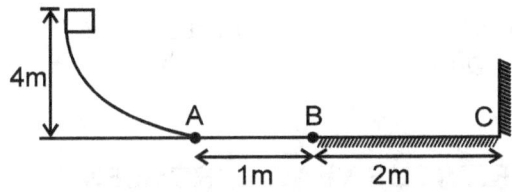

    (A) 29 m (B) 49 m (C) 59 m (D) 109 m

2. A projectile of mass 3m explodes at highest point of its path. It breaks into three equal parts. One part retraces its path, the second one comes to rest. The range of the projectile was 100 m if no explosion would have taken place. The distance of the third part from the point of projection when it finally lands on the ground is –
    (A) 100 m (B) 150 m
    (C) 250 m (D) 300 m

3. A block of mass m is pushed towards a movable wedge of mass nm and height h, with a velocity u. All surfaces are smooth. The minimum value of u for which the block reach the top of the wedge is –

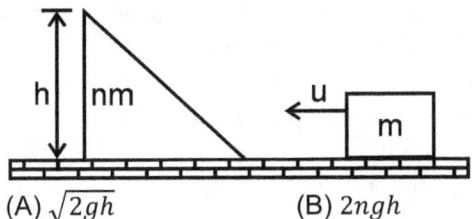

    (A) $\sqrt{2gh}$ (B) $2ngh$
    (C) $\sqrt{2gh\left(1+\frac{1}{n}\right)}$ (D) $\sqrt{2gh\left(1-\frac{1}{n}\right)}$

4. A man of mass m moves with a constant speed on a plank of mass 'M' and length 'L' kept initially at rest on a frictionless horizontal surface, from one end to the other in time 't'. The speed of the plank relative to ground while man is moving, is –
    (A) $\frac{L}{t}\left(\frac{M}{m}\right)$ (B) $\frac{L}{t}\left(\frac{m}{M+m}\right)$
    (C) $\frac{L}{t}\left(\frac{m}{M-m}\right)$ (D) None of these

5. In the figure the block B of mass m starts from rest at the top of a wedge W of mass M. All surfaces are without friction. W can slide on the ground B slides down onto the ground, moves along it with a speed v, has an elastic collision with the wall, and climbs back onto W.

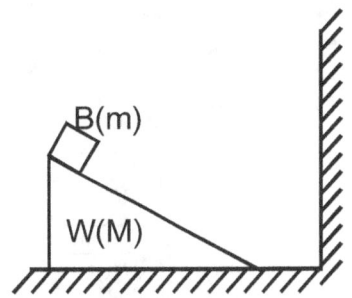

    (A) B will reach the top of W again
    (B) From the beginning, till the collision with the wall, the centre of mass of 'B plus W' does not move horizontally
    (C) After the collision, the centre of mass of 'B plus W' moves with the velocity $\frac{2mv}{m+M}$
    (D) When B reaches its highest position on W, the speed of W is $\frac{2mv}{m+M}$

6. The ring R in the arrangement shown can slide along a smooth, fixed, horizontal rod XY. It is attached to the block B by a light string. The block is released from rest, with the string horizontal.

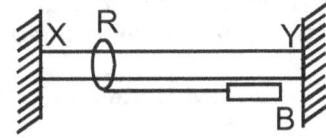

    (A) One point in the string will have only vertical motion
    (B) R and B will always have momenta of the same magnitude.
    (C) When the string becomes vertical, the speed of R and B will be inversely proportionally to their masses
    (D) R will lose contact with the rod at some point

7. A smooth sphere is moving on a horizontal surface with velocity vector $2\hat{i} + 2\hat{j}$ immediately before it hits a vertical wall. The wall is parallel to vector and the coefficient of restitution between the sphere and the wall is $e = \frac{1}{2}$. The velocity vector of the sphere after it hits the wall is:
   (A) $\hat{i} - \hat{j}$ (B) $-\hat{i} + 2\hat{j}$
   (C) $-\hat{i} - \hat{j}$ (D) $2\hat{i} - \hat{j}$

8. A light spring of spring constant k is kept compressed between two blocks of masses m and M on a smooth horizontal surface (figure) When released, the blocks acquire velocities in opposite directions. The spring loses contact with the blocks when it acquires natural length. If the spring was initially compressed through a distance x, find the final speed of mass m.

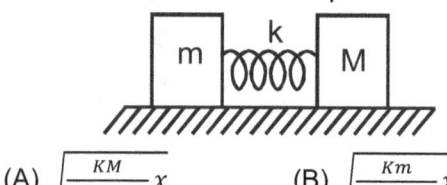

   (A) $\sqrt{\dfrac{KM}{m(M+m)}}x$ (B) $\sqrt{\dfrac{Km}{M(m+M)}}x$
   (C) $\sqrt{\dfrac{KM}{m(M-m)}}x$ (D) $\sqrt{\dfrac{Km}{M(M-m)}}x$

9. Two particles having position vectors $\vec{r}_1 = (3\hat{i} + 5\hat{j})$ meters and $\vec{r}_2 = (-5\hat{i} - 3\hat{j})$ meters are moving with velocities $\vec{v}_1 = (4\hat{i} + 3\hat{j})$ m/s and $\vec{v}_2 = (a\hat{i} + 7\hat{j})$ m/s. If they collide after 2 seconds the value of a is:
   (A) 2   (B) 4   (C) 6   (D) 8

ASSERTION & REASON TYPE QUESTIONS –
Each of the questions given below consist of Statement – I and Statement – II. Use the following Key to choose the appropriate answer.
(A) If both Statement- I and Statement- II are true, and Statement - II is the correct explanation of Statement– I.
(B) If both Statement - I and Statement - II are true but Statement - II is not the correct explanation of Statement – I.
(C) If Statement - I is true but Statement - II is false.
(D) If Statement - I is false but Statement - II is true.

10. **Statement - I:** In an elastic collision of two billiards balls, the kinetic energy is not conserved during the short interval of time of collision between the balls.
    Statement - II: Energy spent against friction does not follow the law of conservation of energy.

••**PASSAGE PROBLEMS**

BIO Momentum and the squirting squid. An interesting use of "rocket power" is that used by cephalopods such as octopi and squids. These animals take in seawater and then squirt it out at high speed. A 2.5-kg squid can expel 0.25 kg of seawater (in a short burst of 0.20 s) with a speed of 600 cm/s.

11. What is the momentum of one squirt of water?
    (A) 1.2 kg m/s in the direction of the squirt
    (B) 1.5 kg m/s in the direction of the squirt
    (C) 12 kg m/s in the direction of the squirt
    (D) 15 kg m/s in the direction of the squirt

12. How much momentum does this give the squid?
    (A) 1.2 kg m/s in the direction of the squirt
    (B) 1.5 kg m/s in the direction of the squirt
    (C) 15 kg m/s in the direction of the squirt
    (D) 1.5 kg m/s in the direction opposite to the squirt

13. What would be the speed of the squid immediately after one squirt?
    (A) 10 cm/s   (B) 20 cm/s
    (C) 40 cm/s   (D) 60 cm/s

14. After three quick squirts (in the same direction) what is the speed of the squid? Ignore the drag on the squid from the ocean water.
    (A) 60 cm/s   (B) 120 cm/s
    (C) 180 cm/s   (D) 200 cm/s

38.4. PREVIOUS YEARS PROBLEMS

38.4.1. ••SECTION-A (JEE MAIN)

1. Two identical particles move towards each other with velocity 2v and v respectively. The velocity of centre of mass is – [2002]
   (A) v (B) $v/3$ (C) $v/2$ (D) zero

2. Consider the following two statements: [2003]
   (i) Linear momentum of a system of particle is zero
   (ii) Kinetic energy of a system of particle is zero
   Then:
   (A) (i) implies (ii) but (ii) does not imply (i)
   (B) (i) does not imply (ii) but (ii) implies (i)
   (C) (i) implies (ii) and (ii) implies (i)
   (D) (i) does not imply (ii) and (ii) does not imply (i)

3. A body A of mass M while falling vertically downwards under gravity breaks into two parts, a body B of mass $\frac{1}{3}M$ and a body C of mass $\frac{2}{3}M$. The centre of mass of bodies B and C taken together shifts compared to that of body A towards [2005]
   (A) depends on height of breaking
   (B) does not shift

(C) body C
(D) body B

4. The block of mass M moving on the frictionless horizontal surface collides with the spring of spring constant k and compresses it by length L. The maximum momentum of the block after collision is
[2005]

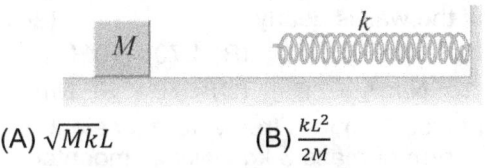

(A) $\sqrt{MkL}$
(B) $\frac{kL^2}{2M}$
(C) zero
(D) $\frac{ML^2}{k}$

5. A mass 'm' moves with a velocity 'v' and collides inelastically with another identical mass. After collision the 1st mass moves with velocity $\frac{v}{\sqrt{3}}$ in a direction perpendicular to the initial direction of motion. Find the speed of the 2nd mass after collision.
[2005]

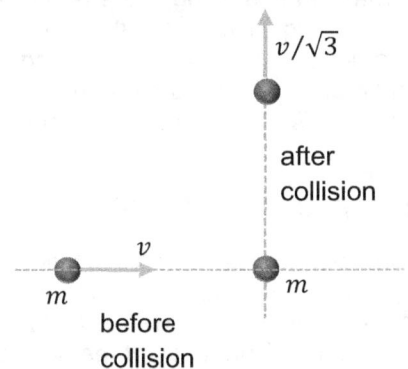

(A) v (B) $\sqrt{3}v$ (C) $\frac{2}{\sqrt{3}}v$ (D) $\frac{v}{\sqrt{3}}$

6. A bomb of mass 16 kg at rest explodes into two pieces of masses 4 kg and 12 kg. The velocity of the 12 kg mass is $4\ ms^{-1}$. The kinetic energy of the other mass is –
[2006]
(A) 192 J   (B) 96 J   (C) 144 J   (D) 288 J

7. Consider a two particle system with particles having masses m1 and m2. If the first particle is pushed towards the centre of mass through a distance d, by what distance should the second particle be moved, so as to keep the centre of mass at the same position –
[2006]
(A) $\frac{m_1}{m_2}d$
(B) d
(C) $\frac{m_2}{m_1}d$
(D) $\frac{m_1}{m_1+m_2}d$

8. A block of mass 0.50 kg is moving with a speed of $2.00\ ms^{-1}$ on a smooth surface. It strikes another mass of 1.00 kg and then they move together as a single body. The energy loss during the collision is
[2008]
(A) 1.00 J   (B) 0.67 J
(C) 0.34 J   (D) 0.16 J

9. A thin rod of length 'L' is lying along the x-axis with its ends at x = 0 and x = L. Its linear density (mass/length) varies with x as $k(x/L)^n$, where n can be zero or any positive number. If the position x_{CM} of the centre of mass of the rod is plotted against 'n', which of the following graphs best approximates the dependence of x_{CM} on n?
[2008]

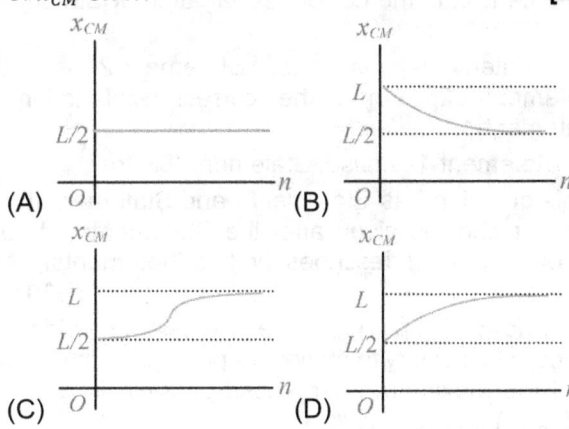

10. Consider a rubber ball freely falling from a height h = 4.9 m onto a horizontal elastic plate. Assume that the duration of collision is negligible and the collision with the plate is totally elastic. Then the velocity as a function of time and the height as function of time will be
[2009]

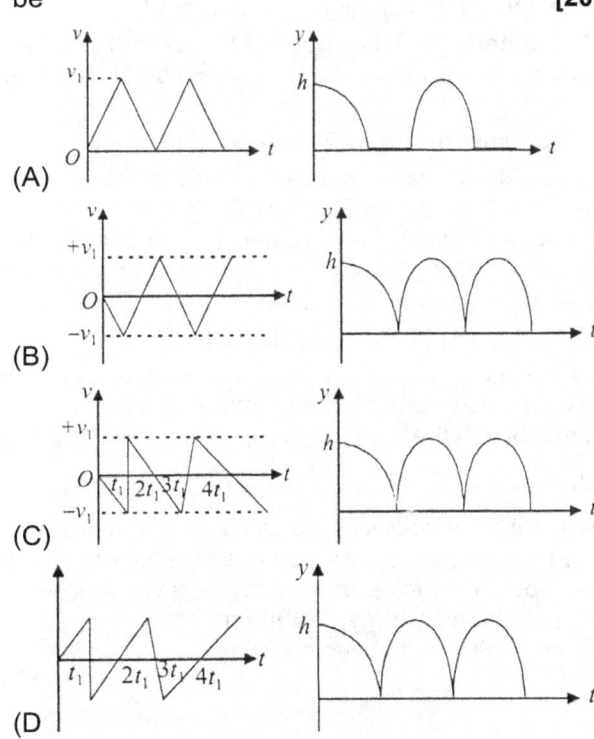

11. This question has Statement-I and Statement-II. Of the four choices given after the Statements, choose the one that best describes the two Statements.
[2010]

Statement-1: Two particles moving in the same direction do not lose all their energy in a completely inelastic collision.

Statement-2: Principle of conservation of momentum holds, true for all kinds of collisions.

(A) Statement-1 is true, Statement-2 is false.
(B) Statement-1 is true, Statement-2 is true; Statement-2 is the correct explanation of Statement-1.
(C) Statement-1 is true, Statement-2 is true; Statement-2 is not the correct explanation of Statement-1.
(D) Statement-1 is false, Statement-2 is true.

12. This question has Statement-I and Statement-II. Of the four choices given after the Statements, choose the one that best describes the two Statements. **[2013]**

Statement-1: A point particle of mass m moving with speed v collides with stationary point particle of mass M. If the maximum energy loss possible is given as $h\left(\frac{1}{2}mv^2\right)$ then $h = \left(\frac{m}{m+M}\right)$

Statement-2: Maximum energy loss occurs when the particles get stuck together as a result of the collision.

(A) Statement-I is false, Statement-II is true.
(B) Statement-I is true, Statement-II is true, Statement-II is a correct explanation of Statement-I.
(C) Statement-I is true, Statement-II is true, Statement-II is not a correct explanation of statement-I.
(D) Statement-I is true, Statement-II is false.

13. A particle of mass m moving in the x direction with speed $2v$ is hit by another particle of mass $2m$ moving in the y direction with speed v. If the collision is perfectly inelastic, the percentage loss in the energy during the collision is close to: **[2015]**
(A) 44 %   (B) 50 %   (C) 56 %   (D) 62 %

14. Distance of the centre of mass of a solid uniform cone from its vertex is z_0. If the radius of its base is R and its height is h the z_0 is equal to: **[2015]**
(A) $\frac{h^2}{4R}$ (B) $\frac{3h}{4}$ (C) $\frac{5h}{8}$ (D) $\frac{5h^2}{8R}$

15. In a collinear collision, a particle with an initial speed v_0 strikes a stationary particle of the same mass. If the final total kinetic energy is 50% greater than the original kinetic energy, the magnitude of the relative velocity between the two particles, after collision, is **[2018]**
(A) $\frac{v_0}{4}$ (B) $\sqrt{2}v_0$ (C) $\frac{v_0}{2}$ (D) $\frac{v_0}{\sqrt{2}}$

16. It is found that if a neutron suffers an elastic collinear collision with deuterium at rest, fractional loss of its energy is p_d; while for its similar collision with carbon nucleus at rest, fractional loss of energy is p_c. The values of p_d and p_c are respectively **[2018]**
(A) (0.89, 0.28)   (B) (0.28, 0.89)
(C) (0, 0)   (D) (0, 1)

17. The mass of a hydrogen molecule is 3.32×10^{-27} kg. If 10^{23} hydrogen molecules strike, per second, a fixed wall of area 2 cm² at an angle of 45° to the normal, and rebound elastically with a speed of 10^3 m/s, then the pressure on the wall is nearly **[2018]**
(A) 2.35×10^3 N/m²   (B) 4.70×10^3 N/m²
(C) 2.35×10^2 N/m²   (D) 4.70×10^2 N/m²

18. A body of mass 1 kg falls freely from a height of 100 m, on a platform of mass 3 kg which is mounted on a spring having spring constant $k = 1.25 \times 10^6$ N/m. The body sticks to the platform and the spring's maximum compression is found to be x. Given that $g = 10\ ms^{-2}$, the value of x will be close to: **[2019]**
(A) 40 cm   (B) 4 cm
(C) 80 cm   (D) 2 cm

19. A particle of mass m is moving in a straight line with momentum p. Starting at time $t = 0$, a force $F = kt$ acts in the same direction on the moving particle during time interval T so that its momentum changes from p to $3p$. Here k is a constant. The value of T is: **[2019]**
(A) $2\sqrt{\frac{k}{p}}$ (B) $2\sqrt{\frac{p}{k}}$ (B) $\sqrt{\frac{2k}{p}}$ (D) $\sqrt{\frac{2p}{k}}$

20. An alpha-particle of mass m suffers 1-deminsional elastic collision with a nucleus at rest of unknown mass. It is scattered directly backwards losing, 64% of its initial kinetic energy. The mass of the nucleus is: **[2019]**
(A) $2\ m$ (B) $3.5\ m$ (C) $1.5\ m$ (D) $4\ m$

21. A particle of mass m is moving with speed $2v$ collides with a mass 2m moving with speed v in the same direction. After collision, the first mass is stopped completely while the second one splits into two particles each of mass m, which move at angle 45° with respect to the original direction. The speed of each of the moving particle will be: **[2019]**
(A) $v/2\sqrt{2}$ (B) $2\sqrt{2}v$
(C) $\sqrt{2}v$ (D) $v/\sqrt{2}$

22. If 10^{22} gas molecules each of mass 10^{-26} kg, collide with a surface (perpendicular to it) elastically per second over an area $1 m^2$ with a speed $10^4\ m/s$, the pressure exerted by the gas molecules will be of the order of: **[2019]**
(A) $2\ N/m^2$ (B) $10^3\ N/m^2$
(C) $10^4\ N/m^2$ (D) $10^{16}\ N/m^2$

23. A body of mass m moving with velocity 'v' collides with shown masses respectively. Find loss in KE after the last collision. Consider all collision completely inelastically? **[2020]**

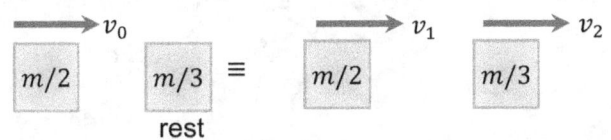

(A) 85.5 (B) 90.2 (C) 93.75 (D) 88.5

24. A Body of mass $m/2$ moving with velocity v_0 collides elastically with another mass of $m/3$. Find % change in KE of first body? **[2020]**

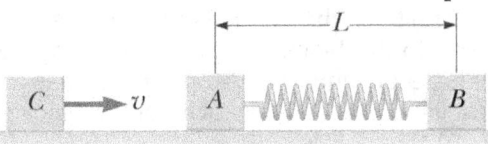

(A) 32% (B) 96% (C) 34% (D) 80 %

25. A bowling machine projects a ball of mass 0.15 kg in upward direction. If ball displaced along bowling machine 0.2 m and released. After the released from bowling machine ball attain 20 m height then find the force exerted by bowling machine on the ball. **[2020]**

(A) 145.5 N (B) 165.5 N
(C) 175.5 N} (D) 151.5 N

38.4.2. ●●●SECTION - B [ADVANCED]

1. An isolated particle of mass m is moving in the horizontal plane (x – y), along the x-axis, at a certain height above the ground. It suddenly explodes into two fragments of masses m/4 and 3m/4. An instant later, the smaller fragment is at y = +15 cm. The larger fragment at this instant is at **[1997]**
   (A) $y = -5\, cm$ (B) $y = +20\, cm$
   (C) $y = +5\, cm$ (D) $y = -20\, cm$

2. Two blocks A and B, each of mass m are connected by a massless spring of natural length L and spring constant K. The blocks are initially resting on a smooth horizontal floor with the spring at its natural length, as shown in the adjoining figure. A third identical block C also of mass m, moves on the floor with a speed v along the line joining A and B and collides with A. Then, **[1993]**

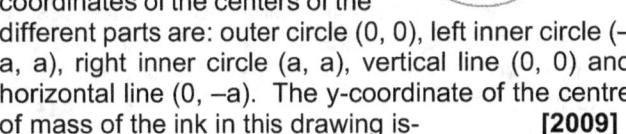

(A) The kinetic energy of the A–B system, at maximum compression of the spring, is zero
(B) The kinetic energy of the A–B system, at maximum compression of the spring, is mv2/4
(C) The maximum compression of the spring is $v\sqrt{m/K}$
(D) None of these

3. Two blocks of masses 10 kg and 4 kg are connected by a spring of negligible mass and placed on a frictionless horizontal surface. An impulse gives a velocity of 14 m/s to the heavier block in the direction of the lighter block. The velocity of the centre of mass is **[2002]**
   (A) 30 m/s (B) 20 m/s
   (C) 10 m/s (D) 5 m/s

4. **Statement Type Question** **[2007]**
   Statement-1 In an elastic collision between two bodies, the relative speed of the bodies after collision is equal to the relative speed before the collision.
   Statement-2 In an elastic collision, the linear momentum of the system is conserved
   (A) Statement-1 is True, Statement-2 is True; Statement-2 is a correct explanation for Statement-1
   (B) Statement-1 is True, Statement-2 is True; Statement-2 is not a correct explanation for Statement-1
   (C) Statement-1 is True, Statement-2 is False
   (D) Statement-1 is False, Statement-2 is True.

5. Two balls, having linear momenta $\vec{p}_1 = p\hat{i}$ and $\vec{p}_2 = -p\hat{i}$, undergo a collision in free space. There is no external force acting on the balls. Let $\vec{p}'_1$ and $\vec{p}'_2$ be their final momenta. The following option(s) is (are) NOT ALLOWED for any non-zero value of p, a_1, a_2, b_1, b_2, c_1 and c_2. **[2007]**
   (A) $\vec{p}'_1 = a_1\hat{i} + b_1\hat{j} + c_1\hat{k};\ \vec{p}'_2 = a_2\hat{i} + b_2\hat{j}$
   (B) $\vec{p}'_1 = c_1\hat{k};\quad \vec{p}'_2 = c_2\hat{k}$
   (C) $\vec{p}'_1 = a_1\hat{i} + b_1\hat{j} + c_1\hat{k}\ ;\ \vec{p}'_2 = a_2\hat{i} + b_2\hat{j} - c_1\hat{k}$
   (D) $\vec{p}'_1 = a_1\hat{i} + b_1\hat{j};\ \vec{p}'_2 = a_2\hat{i} + b_1\hat{j}$

6. Look at the drawing given in the figure which has been drawn with ink of uniform line-thickness. The mass of ink used to draw each of the two inner circles, and each of the two lines segments is m. The mass of the ink used to draw the outer circle is 6m. The coordinates of the centers of the different parts are: outer circle (0, 0), left inner circle (–a, a), right inner circle (a, a), vertical line (0, 0) and horizontal line (0, –a). The y-coordinate of the centre of mass of the ink in this drawing is- **[2009]**
   (A) $\frac{a}{10}$ (B) $\frac{a}{8}$ (C) $\frac{a}{12}$ (D) $\frac{a}{3}$

7. Two small particles of equal masses start moving in opposite directions from a point A in a horizontal circular orbit. Their tangential velocities are v and $2v$, respectively, as shown in the figure. Between collisions, the particles move with constant speeds. After making how many collisions, other than that at A, these two particles will again reach the point A? **[2009]**

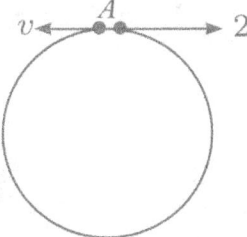

(A) 4   (B) 3   (C) 2   (D) 1

8. If the resultant of all the external forces acting on a system of particles is zero, then from an inertial frame, one can surely say that **[2009]**
   (A) linear momentum of the system does not change in time
   (B) kinetic energy of the system does not change in time
   (C) angular momentum of the system does not change in time
   (D) potential energy of the system does not change in time

9. A point mass of 1 kg collides with a stationary point mass of 5 kg. After their collision, the 1 kg mass reverses its direction and moves with a speed of 2 ms^{-1}. Which of the following statement (s) is (are) correct for the system of these two masses? **[2010]**
   (A) Total momentum of the system is 3 kg ms^{-1}
   (B) Momentum of 5 kg mass after collision is $4\ kg\ ms^{-1}$
   (C) Kinetic energy of the center of mass is 0.75 J
   (D) Total kinetic energy of the system is 4 J

10. A particle of mass m is projected from the ground with an initial speed u_0 at an angle α with the horizontal. At the highest point of its trajectory, it makes a completely inelastic collision with another identical particle, which was thrown vertically upward from the ground with the same initial speed u_0. The angle that the composite system makes with the horizontal immediately after the collision is **[2013]**
    (A) $\frac{\pi}{4}$
    (B) $\frac{\pi}{4} + \alpha$
    (C) $\frac{\pi}{2} - \alpha$
    (D) $\frac{\pi}{2}$

11. A tennis ball is dropped on a horizontal smooth surface. It bounces back to its original position after hitting the surface. The force on the ball during the collision is proportional to the length of compression of the ball. Which one of the following sketches describes the variation of its kinetic energy K with time t most appropriately? The figures are only illustrative and not to the scale. **[2014]**

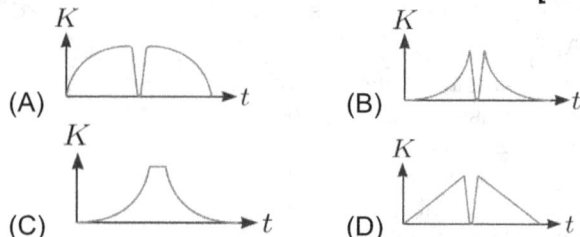

12. A block of mass M has a circular cut with a frictionless surface as shown. The block rests on the horizontal frictionless surface of a fixed table. Initially the right edge of the block is at $x = 0$, in a *co-ordinate system fixed to the table*. A point mass m is released from rest at the topmost point of the path as shown and it slides down. When the mass loses contact with the block, its position is x and the velocity is v. At that instant, which of the following options is/are correct? **[2017]**

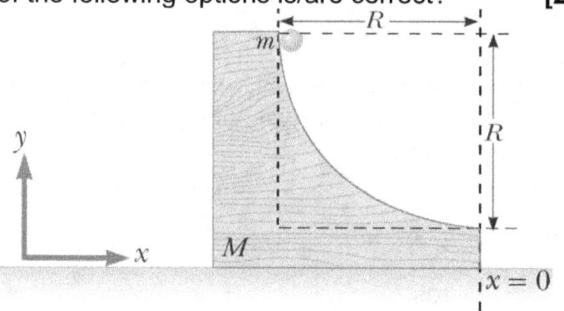

(A) The x component of displacement of the centre of mass of the block M is: $-\frac{mR}{M+m}$
(B) The position of the point mass is: $-\sqrt{2}\frac{mR}{M+m}$
(C) The velocity of the point mass m is: $v = \sqrt{\frac{2gR}{1+\frac{m}{M}}}$
(D) The velocity of the block M is: $v = -\frac{m}{M}\sqrt{2gR}$

13. Consider regular polygons with number of sides $n = 3, 4, 5, \ldots$ as shown in the figure. The center of mass of all the polygons is at height h from the ground. They roll on a horizontal surface about the leading vertex without slipping and sliding as depicted. The maximum increase in height of the locus of the center of mass for each polygon is Δ. Then Δ depends on n and h as: **[2017]**

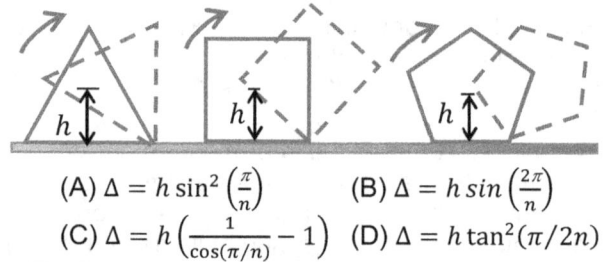

(A) $\Delta = h \sin^2\left(\frac{\pi}{n}\right)$ (B) $\Delta = h \sin\left(\frac{2\pi}{n}\right)$
(C) $\Delta = h \left(\frac{1}{\cos(\pi/n)} - 1\right)$ (D) $\Delta = h \tan^2(\pi/2n)$

INTEGER ANSWER TYPE QUESTIONS

14. A block of mass 0.18 kg is attached to a spring of force-constant 2 N/m. The coefficient of friction between the block and the floor is 0.1. Initially the block is at rest and the spring is un-stretched. An impulse is given to the block as shown in the figure. The block slides a distance of 0.06 m and comes to rest for the first time. The initial velocity of the block in m/s is $v = N/10$. Then, N is **[2011]**

15. A bob of mass m, suspended by a string of length l_1 is given a minimum velocity required to complete a full circle in the vertical plane. At the highest point, it collides elastically with another bob of mass m suspended by a string of length l_2, which is initially at rest. Both the strings are mass-less and inextensible. If the second bob, after collision acquires the minimum

speed required to complete a full circle in the vertical plane, the ratio $\frac{l_1}{l_2}$ is [2013]

COMPREHENSION TYPE QUESTIONS PASSAGE
A small block of mass M moves on a frictionless surface of an inclined plane, as shown in figure. The angle of the incline suddenly changes from 60° to 30° at point B. The block is initially at rest at A. Assume that collisions between the block and the incline are totally inelastic ($g = 10$ m/s^{-2}). [2008]

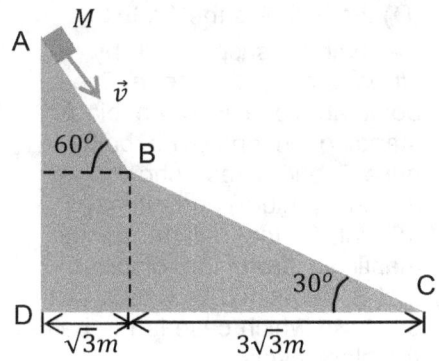

16. The speed of the block at point B immediately after it strikes the second incline is
    (A) $\sqrt{60}$ m/s (B) $\sqrt{45}$ m/s
    (C) $\sqrt{30}$ m/s (D) $\sqrt{15}$ m/s

17. The speed of the block at point C, immediately before it leaves the second incline is
    (A) $\sqrt{120}$ m/s (B) $\sqrt{105}$ m/s
    (C) $\sqrt{90}$ m/s   (D) $\sqrt{75}$ m/s

18. If collision between the block and the incline is completely elastic, then the vertical (upward)
    (A) $\sqrt{30}$ m/s   (B) $\sqrt{15}$ m/s
    (C) 0 m/s   (D) $-\sqrt{15}$ m/s

19. A solid horizontal surface is covered with a thin layer of oil. A rectangular block of mass $m = 0.4$ kg is at rest on this surface. An impulse of 1.0 N s is applied to the block at time to t = 0 so that it starts moving along the x-axis with a velocity $v(t) = v_0 e^{-t/\tau}$, where v_0 is a constant and $\tau = 4\ s$. The displacement of the block, in meters, at $t = \tau$ is (Take $e^{-1} = 0.37$) [2018]

20. A ball is projected from the ground at an angle of 45° with the horizontal surface. It reaches a maximum height of 120 m and returns to the ground. Upon hitting the ground for the first time, it loses half of its kinetic energy. Immediately after the bounce, the velocity of the ball makes an angle of 30° with the horizontal surface. The maximum height it reaches after the bounce, in meters, is [2018]

21. In the List-I below, four different paths of a particle are given as functions of time. In these functions, α and β are positive constants of appropriate dimensions and $\alpha \neq \beta$. In each case, the force acting on the particle is either zero or conservative. In List–II, five physical quantities of the particle are mentioned; $\vec{p}$ is the linear momentum $\vec{L}$ is the angular momentum about the origin, K is the kinetic energy, U is the potential energy and E is the total energy. Match each path in List–I with those quantities in List–II, which are conserved for that path [2018]

List-I	List-II
P. $\vec{r}(t) = \alpha t \hat{i} + \beta t \hat{j}$	1. $\vec{p}$
Q. $\vec{r}(t) = \alpha \cos \omega t \hat{i} + \beta \sin \omega t \hat{j}$	2. $\vec{L}$
R. $\vec{r}(t) = \alpha (\cos \omega t \hat{i} + \sin \omega t \hat{j})$	3. K
S. $\vec{r}(t) = \alpha t \hat{i} + \frac{\beta}{2} t^2 \hat{j}$	4. U
	5. E

(A) P → 1, 2, 3, 4, 5; Q → 2,5; R → 2, 3, 4, 5; S → 5
(B) P → 1, 2, 3, 4, 5; Q → 3,5; R → 2, 3, 4, 5; S → 2, 5
(C) P → 2, 3, 4; Q → 5; R → 1, 2, 4; S → 2, 5
(D) P → 1, 2, 3, 5; Q → 2, 5; R → 2, 3, 4, 5; S → 2, 5

22. A spring-block system is resting on a frictionless floor as shown in the figure. The spring constant is $2.0\ N\ m^{-1}$ and the mass of the block is 2.0 kg. Ignore the mass of the spring. Initially the spring is in an unstretched condition. Another block of mass 1.0 kg moving with a speed of 2.0 m s^{-1} collides elastically with the first block. The collision is such that the 2.0 kg block does not hit the wall. The distance, in meters, between the two blocks when the spring returns to its unstretched position for the first time after the collision is. [2018]

38.5. MISCELLANEOUS QUESTIONS

1. ●●A fireworks projectile is launched upward at an angle above a large flat plane. When the projectile reaches the top of its flight, at a height of h above a point that is a horizontal distance D from where it was launched, the projectile explodes into two equal pieces. One piece reverses its velocity and travels directly back to the launch point. How far from the launch point does the other piece land?
   (A) D (B) 2D (C) 3D (D) 4D

2. ●For a totally elastic collision between two objects, which of the following statements is (are) true?
   (A) The total mechanical energy is conserved.
   (B) The total kinetic energy is conserved.
   (C) The total momentum is conserved.
   (D) The momentum of each object is conserved.

3. ●For a totally inelastic collision between two objects, which of the following statements is (are) true?

(A) The total mechanical energy is conserved.
(B) The total kinetic energy is conserved.
(C) The total momentum is conserved.
(D) The total momentum after the collision is always zero.

4. ●●A ballistic pendulum is used to measure the speed of a bullet shot from a gun. The mass of the bullet is 50.0 g, and the mass of the block is 20.0 kg. When the bullet strikes the block, the combined mass rises a vertical distance of 5.00 cm. What was the speed of the bullet as it struck the block?
(A) 397 m/s   (B) 426 m/s
(C) 457 m/s   (D) 479 m/s

5. ●If two particles have equal momenta, are their kinetic energies equal?
(A) yes, always
(B) no, never
(C) no, except when their speeds are the same
(D) yes, as long as they move along parallel lines.

6. ●If two particles have equal kinetic energies, are their momenta equal?
(A) yes, always
(B) no, never
(C) yes, as long as their masses are equal
(D) yes, if both their masses and directions of motion are the same

7. ●A 10.0-g bullet is fired into a 200-g block of wood at rest on a horizontal surface. After impact, the block slides 8.00 m before coming to rest. If the coefficient of friction between the block and the surface is 0.400, what is the speed of the bullet before impact?
(A) 106 m/s   (B) 166 m/s
(C) 226 m/s   (D) 286 m/s

8. ●●Two particles of different mass start from rest. The same net force acts on both of them as they move over equal distances. How do their final kinetic energies compare?
(A) The particle of larger mass has more kinetic energy.
(B) The particle of smaller mass has more kinetic energy.
(C) The particles have equal kinetic energies.
(D) Either particle might have more kinetic energy.

9. ●Two particles of different mass start from rest. The same net force acts on both of them as they move over equal distances. How do the magnitudes of their final momenta compare?
(A) The particle of larger mass has more momentum.
(B) The particle of smaller mass has more momentum.
(C) The particles have equal momenta.
(D) Either particle might have more momentum.

10. ●A basketball is tossed up into the air, falls freely, and bounces from the wooden floor. From the moment after the player releases it until the ball reaches the top of its bounce, what is the smallest system for which momentum is conserved?
(A) the ball
(B) the ball plus player
(C) the ball plus floor
(D) the ball plus the Earth

11. ●A ball is suspended by a string that is tied to a fixed point above a wooden block standing on end. The ball is pulled back (as shown in adjoining figure) and released. In trial A, the ball rebounds elastically from the block. In trial B, two-sided tape causes the ball to stick to the block. In which case is the ball more likely to knock the block over?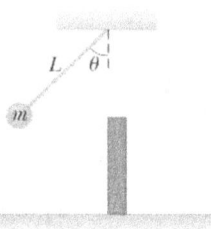
(A) It is more likely in trial A.
(B) It is more likely in trial B.
(C) It makes no difference.
(D) It could be either case, depending on other factors.

12. ●A car of mass m traveling at speed v crashes into the rear of a truck of mass 2m that is at rest and in neutral at an intersection. If the collision is perfectly inelastic, what is the speed of the combined car and truck after the collision?
(A) v (B) $v/2$ (C) $v/3$ (D) $2v$

13. ●●A hockey puck of mass m moves at speed v on ice. A hockey player deflects the puck such that its speed stays the same, but the puck is now moving at 60° with respect to its incident direction. What is the magnitude of the impulse?
(A) zero   (B) $mv \sin 30°$
(C) mv (D) $mv \cos 30°$

14. ●●Suppose Earth were the only planet in the solar system. As Earth orbited the Sun,
(A) the Sun would remain stationary
(B) the Sun would always move in the same direction as Earth
(C) the Earth–Sun centre of mass would move back and forth
(D) both Earth and the Sun would orbit the Earth–Sun centre of mass.

15. ●●If one of the stars in a binary star system explodes, and the intact star absorbs a good portion of the material from the first star, then
(A) the centre of mass of the exploding star is not

affected by the explosion

(B) the centre of mass of the two-star system is not affected by the explosion

(C) the centre of mass of the two-star system will move closer to the intact star as the material from the exploding star gets absorbed by the intact star

(D) None of these

16. ●●A cannon fires a probe into the Moon such that the probe leaves Earth's surface at 90° to the surface. Which of the following statements is true?
(A) Earth will not be affected at all.
(B) Earth will recoil slightly.
(C) Since the force of gravity is involved, momentum is not conserved.
(D) None of the above are true

17. ●A rocket is fired vertically upward from Earth's surface. If you chose a system in which momentum is conserved during the process, the most correct system to pick would be
(A) rocket + fuel/exhaust
(B) rocket + fuel/exhaust + atmosphere
(C) rocket + fuel/exhaust + atmosphere + Earth
(D) rocket + fuel/exhaust + atmosphere + Milky Way galaxy

18. ●Two particles A and B of equal mass are located at some distance from each other. Particle A is at rest while B moves away from A at speed v. What happens to the center of mass of the system of two particles?
(A) It does not move.
(B) It moves with a speed v away from A.
(C) It moves with a speed v toward A.
(D) It moves with a speed $\frac{1}{2}v$ away from A.
(D) It moves with a speed $\frac{1}{2}v$ toward A.

19. ●Two uniform spheres with equal mass per unit volume are in contact with one another. The mass of sphere A is five times that of sphere B. The center of mass of the system is
(A) at the point where A and B touch.
(B) inside sphere B somewhere on the line joining the centers of A and B.
(C) inside sphere A somewhere on the line joining the centers.
(D) at the center of sphere A.

20. ●●An object at rest suddenly explodes into three parts of equal mass. Two of the parts move away at right angles to each other and with equal speeds v. What is the velocity of the third part just after the explosion?

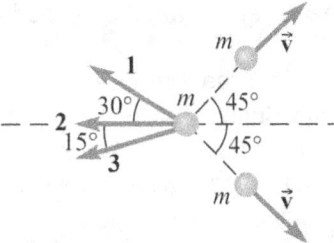

(A) Direction of vector 1 and magnitude $2v$
(B) Direction of vector 2 and magnitude $\sqrt{2}v$
(C) Direction of vector 3 and magnitude $\frac{1}{\sqrt{2}}v$
(D) Direction of vector 2 and magnitude $\frac{1}{\sqrt{2}}v$

21. ●●An object of mass m drops from rest a little above the Earth's surface for a time t. Ignore air resistance. After time t the magnitude of its momentum is
(A) mgt^2 (B) mgt
(C) $mg\sqrt{t}$ (D) $\sqrt{mgt}$

●●**Multiple-Choice Questions 22–27** refer to a situation in which a golf ball is projected straight upward in the +y-direction. Ignore air resistance. The answer choices are found in the figure.

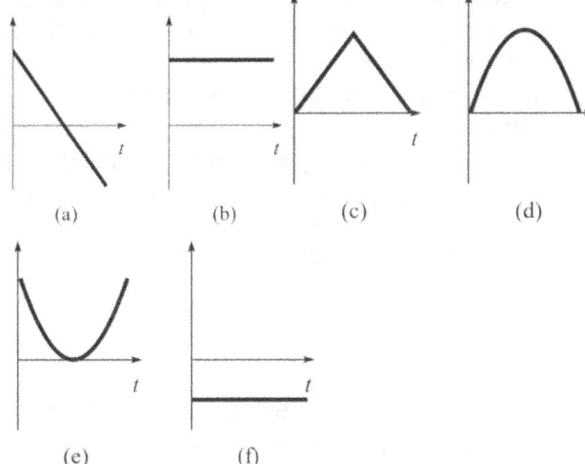

22. Which graph shows the acceleration a_y of the ball as a function of time?
23. Which graph shows the vertical position y of the ball as a function of time?
24. Which graph shows the momentum p_y of the ball as a function of time?
25. Which graph shows the kinetic energy of the ball as a function of time?
26. Which graph shows the potential energy of the ball as a function of time?
27. Which graph shows the total energy of the ball as a function of time?
28. The gravitational force that Earth exerts on an object causes an impulse of +10 N.s in one experiment and

+1 N. s on the same object in another experiment. How can this be?
(A) The mass of the object changed
(B) The time intervals during which the force was exerted are different.
(C) The magnitudes of the force were different
(D) Not possible

29. •A ball of mass 0.18 kg moving with speed collides head-on with an identical stationary ball. (Notice that we do not know the type of collision.) Which of the following quantities can be calculated from this information alone?
(A) The force each ball exerts on the other.
(B) The velocity of each ball after the collision.
(C) Total kinetic energy of both balls after the collision.
(D) Total momentum of both balls after the collision.

30. •In which of the following collisions would you expect the kinetic energy to be conserved? (There may be more than one correct choice.)
(A) A bullet passes through a block of wood.
(B) Two bull elk charge each other and lock horns.
(C) Two asteroids collide by a glancing blow, but do not actually hit each other, and their only interaction is through gravity.
(D) Two cars with spring-like bumpers collide at fairly low speeds.

31. ••Cart A, of mass 1 kg, approaches and collides with cart B, which has a mass of 4 kg and is initially at rest. (See adjoining figure) When the springs have reached their maximum compression,

(A) cart A has come to rest relative to the ground.
(B) both carts have the same velocity.
(C) both carts have the same momentum.
(D) all the initial kinetic energy of cart A has been converted to elastic potential energy.

32. •Two lumps of clay having equal masses and speeds, but traveling in opposite directions on a frictionless horizontal surface, collide and stick together. Which of the following statements about this system of lumps must be true? (There may be more than one correct choice.)
(A) The momentum of the system is conserved during the collision.
(B) The kinetic energy of the system is conserved during the collision.
(C) The two masses lose all their kinetic energy during the collision.

(D) The velocity of the center of mass of the system is the same after the collision as it was before the collision.

33. •A heavy rifle initially at rest fires a light bullet. Which of the following statements about these objects is true? (There may be more than one correct choice.)
(A) The bullet and rifle both gain the same magnitude of momentum.
(B) The bullet and rifle are both acted upon by the same average force during the firing.
(C) The bullet and rifle both have the same acceleration during the firing.
(D) The bullet and the rifle gain the same amount of kinetic energy.

34. •You drop an egg from rest with no air resistance. As it falls,
(A) only its momentum is conserved.
(B) only its kinetic energy is conserved.
(C) both its momentum and its mechanical energy are conserved.
(D) its mechanical energy is conserved, but its momentum is not conserved.

35. •A ball is dropped from a height h on the ground. If the coefficient of restitution is e, the height to which the ball goes up after it rebounds for the nth time is.
(A) he^{2n} (B) he^n (C) $\frac{e^{2n}}{h}$ (D) $\frac{h}{e^{2n}}$

36. ••A particle of mass m moving with velocity u makes an elastic one-dimensional collision with a stationary particle of mass m. They are in contact for a very brief time T. Their force of interaction increases from zero to F_0 linearly in time T/2 and decreases linearly to zero in further time T/2. The magnitude of F_0 is –

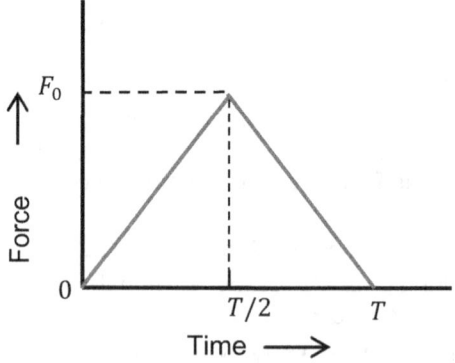

(A) mu/T (B) $2mu/T$
(C) $mu/2T$ (D) None of these

39. ANSWER KEYS AND SOLUTIONS

39.1. CHECKPOINT 1

1. $v_B = 0.783$ m/s, $v_A = -0.675$ m/s,

2. (a) 26.00 $\hat{i}$ m/s (b) 8.40 J (c) The original energy is in the spring. (d) A force had to be exerted over a displacement to compress the spring, transferring energy into it by work. The cord exerts force, but over no displacement. (e) System momentum is conserved with the value zero. (f) The forces on the two blocks are internal forces, which cannot change the momentum of the system; the system is isolated. (g) Even though there is motion afterward, the final momenta are of equal magnitude in opposite directions, so the final momentum of the system is still zero. 3. 6.9 m/s

39.2. CHECKPOINT 2

1. (c) no difference, 2. (a) 12.0 $\hat{i}$ N.s (b) 4.80 $\hat{i}$ m/s (c) 2.80 $\hat{i}$ m/s (d) 2.40 $\hat{i}$ N, 3. $v_0 = 12.5\ ms^{-1}$

39.3. CHECKPOINT 3

1. μV_0, 2. 0, 3. $x = 6.18\ m$, 4. $v = \sqrt{gl \sin\theta}$, 5. $v = 7.90$ m/s, 6. $a = \frac{F}{\rho x} - \mu_k g - \frac{v^2}{x}$, 7. $m = 57.6 \times 10^3$ kg, $\frac{dm}{dt} = 1216$ kg/s, $a = 0.104\ m/s^2$, 8. $F = \rho v^2$, 9. $F = 20.4\ N$, 11. 16.5 N

39.4. CHECKPOINT 4

1. m = 10.86 Mg, 2. $m = m_0 e^{-\frac{a+g}{u}t}$, 3. 15.0 N in the direction of the initial velocity of the exiting water stream.

39.5. CHECKPOINT 5

1. $3\rho g x$, 2. $N = \rho g x + \rho v^2$, (3) $v = \sqrt{\frac{2}{3}g\left(\frac{y^3-h^3}{y^2}\right)}$ 4. (a) $v_1 = \sqrt{2gh \ln\frac{L}{h}}$ (b) $v_2 = \sqrt{2gh\left[1 + \ln\frac{L}{h}\right]}$ (c) $Q = \rho g h \left[L - \frac{h}{2}\right]$, 5. $F = \frac{1}{2}\rho\left[\frac{v^2}{2} + gx\right]$, $R = \frac{\rho v^2}{4} + \rho g \left(L - \frac{x}{2}\right)$, $T_1 = \frac{\rho v^2}{4}$, 6. $F = \frac{1}{2}\rho[v^2 + gx]$, $R = \rho g \left(L - \frac{x}{2}\right)$, $T_1 = \frac{\rho v^2}{2}$, $Q = \frac{\rho x}{4}(v^2 + gx)$, 7. $\Delta Q = \rho g r^2$, 8. (a) $a = \frac{g}{L}x$, (b) $T = \rho g x \left(1 - \frac{x}{L}\right)$ (c) $v = \sqrt{gL}$, 9. (a) $a = \frac{h}{H}g$, (b) $v = h\sqrt{\frac{g}{H}}$, (c) $R = 2\rho g \left(H - \frac{2h^2}{H}\right)$

39.6. CHECKPOINT 6

1. (a) $v_{wedge} = -0.667\ m/s$, (b) $h = 0.952\ m$, 2. (a) $v_b = -mv/(m + M)$ (b) When the man stops so does the balloon. 3. (a) Assume the car is massless. Then moving the cannon balls is moving the center of mass, unless the cannon balls don't move but instead the car does. How far? L. (b) Once the cannonballs stop moving so does the rail car, 4. (a) 6.9 m/s (b) 30°(c) 6.9 m/s (d) −30° (e) 2.0 m/s (f) −180°, 5. (a) 25 mm; (b) 26 mm; (c) down; (d) $1.6 \times 10^{-2}\ m/s^2$, 6. $\frac{1}{4}m(v_1 - v_2)^2$

39.7. CHECKPOINT 7

1. Last ball will move with the same speed.

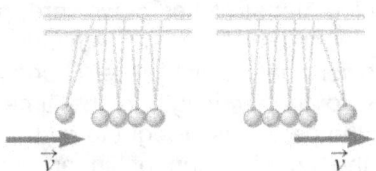

39.8. CHECKPOINT 8

1. 4.85 m/s, 2. $v = \frac{4M}{m}\sqrt{gl}$, 3. 0.556 m, 4. (a) $\sqrt{\frac{F(2d-l)}{2m}}$ (b) $\frac{Fl}{2}$, 5. (a) Momentum of the bullet-block system is conserved in the collision, so you can relate the speed of the block and bullet right after the collision to the initial speed of the bullet. Then, you can use conservation of mechanical energy for the bullet-block-Earth system to relate the speed after the collision to the maximum height. (b) $v_i = \frac{m+M}{m}\sqrt{2gh}$. 6. (a) Momentum of the bullet-block system is conserved in the collision, so you can relate the speed of the block and bullet right after the collision to the initial speed of the bullet. Then, you can use conservation of mechanical energy for the bullet-block-Earth system to relate the speed after the collision to the maximum height. (b) $v_i = 521$ m/s, 7. m_1: 13.9 m, m_2: 0.556 m, 8. 143 m/s, 9. (a) d = 2.22 m (b) d = 0.556 m, 10. (a) −3.54 m/s. (b) 1.77 m (c) 3.54×10^4 N, (d) No. The spring exerts a force on the system during the firing. The force represents an impulse, so the momentum of the system is not conserved in the horizontal direction. Consider the vertical direction. There are two vertical forces on the system: the normal force from the ground and the gravitational force. During the firing, the normal force is larger than the gravitational force. Therefore, there is a net impulse on the system in the upward direction. The impulse accounts for the initial vertical momentum component of the projectile.

39.9. CONCEPTUAL QUESTIONS

1. The likelihood of injury resulting from jumping from a second-floor window is primarily determined by the average force acting to decelerate the body.
(a) The deceleration time interval for a person landing stiff legged on pavement is very short. The impulse

momentum theorem tells us that the average force acting on the person's feet must therefore be very large such a person is likely to incur injuries.
(b) Jumping into a privet hedge increases the time interval over which the body decelerates. This decreases the average force on the person's limbs and therefore decreases the likelihood of injury.
(c) Jumping into a firefighter's net is the best option of the three. The net stretches downward, gradually bringing the person to rest. Additionally, the firefighters lower the net with their hands as the person lands to further lengthen the time interval during which the person is brought to rest.

2. The law of the conservation of linear momentum states that in the absence of external interactions, the linear momentum of a closed system is constant. Floating in free space, the astronaut and the wrench form a closed system free from interactions with other bodies. If the astronaut throws the wrench in the direction opposite the ship, conservation of momentum dictates that he must in turn move toward the ship.

3. In an elastic collision between the hammer and nail, the kinetic energy of the system is conserved while in a perfectly inelastic collision, the greatest percentage of the kinetic energy is lost. The energy lost by the system in a perfectly inelastic collision is used to do the work required to bring the hammer and nail together. In an elastic collision, this work is available to drive the nail into the wood—the total work available to drive the nail is therefore greater for an elastic collision. Thus, for equal applied forces, the hammer will drive the nail further into the wood if the collision is elastic.

4. First law: The momentum of an object is constant unless acted upon by an external force. Second law: The net force acting on an object is equal to the rate of change of the object's momentum. Third law: When two objects interact, the changes in momentum that each imparts to the other are equal in magnitude and opposite in direction.

5. The frictional force of the road on the tires supplies the external force to change the bicycle's momentum. Changes in the bicycle's kinetic energy do not require an external force. For example, the rider could throw her helmet away hard, increasing both her and the helmet's speed. The kinetic energy of the system (bicycle, rider, and helmet) would increase, while the momentum would remain the same. Note that the work-energy theorem (total work done equals change in kinetic energy) cannot be used here, because the internal structure of the system cannot be ignored.

6. Raju is correct. In a fall, the two ropes will experience the same change in momentum. The change in momentum to zero in the stretching rope will take longer time than a stiff rope. As change in momentum is equal to the product of force and time, hence a lesser force will be created in the stretching rope, thereby keeping the rope in better condition and also comfort the faller.

7. The momentum of the keys increases as they fall because a net force acts on them. The momentum of the universe is unchanged because an equal and opposite force acts on the Earth.

8. If the kinetic energy is zero the speeds must be zero as well. This means that the momentum is zero.

9. Yes, in much the same way that a propeller in water can power a speedboat.

10. When a heavy object and a light object collide they exert equal and opposite forces on one another. Since the light object has less mass, its acceleration is greater. This can result in more severe injuries for the light vehicle.

11. (a) The force due to braking—which ultimately comes from friction with the road—reduces the momentum of the car. The momentum lost by the car does not simply disappear, however. Instead, it shows up as an increase in the momentum of the Earth. (b) As with braking, the ultimate source of the force accelerating the car is the force of static friction between the tires and the road.

12. (a) Yes. Suppose two objects have momenta of equal magnitude. If these objects collide in a head-on, completely inelastic collision, they will be at rest after the collision. In this case, all of the initial kinetic energy is converted to other forms of energy. (b) No. In order for its momentum to change, an external force must act on the system. We are given, however, that the system is isolated. Therefore, the only forces acting on it are internal forces.

13. No. Any collision between cars will be at least partially inelastic, due to denting, sound production, heating, and other effects.

14. The center of mass of the hourglass starts at rest in the upper half of the glass and ends up at rest in the lower half. Therefore, the center of mass accelerates downward when the sand begins to fall—to get it moving downward—and then accelerates upward when most of the sand has fallen—to bring it to rest again. It follows from equation, $\sum F = Ma_{CM}$, that the weight read by the scale is less than Mg when the sand begins falling, but is greater than Mg when most of the sand has fallen.

15. (a) Assuming a very thin base, we conclude that the center of mass of the glass is at its geometric center of the glass. (b) In the early stages of filling, the center of mass is below the center of the glass. When the glass is practically full, the center of mass is again at the geometric center of the glass. Thus, as water is added, the center of mass first moves downward, then turns around and moves back upward to its initial position.

16. Your center of mass is somewhere directly above the area of contact between your foot and the ground.
17. As this jumper clears the bar, a significant portion of his body extends below the bar due to the extreme arching of his back. Just as the center of mass of a donut can lie outside the donut, the center of mass of the jumper can be outside his body. In extreme cases, the center of mass can even be below the bar at all times during the jump.

39.10. PROBLEMS

Q.	1	2	3
ANS	4 Ns	1.0×10^3 N	8.0×10^2 N, left
Q.	4	5	6
ANS	80 m/s to the left	(a) 1.5×10^4 N.s (b) 30 s, 110 m/s	1.5×10^3 N
Q.	7	8	9
ANS	0.20 s	6.0 m/s	(a) 1.0×10^3 kg.m/s (b) 1.3×10^4 N
Q.	10	11	12
ANS	2.1 kg m/s, to the left	$5.0 \times 10^2 kg$	2.4 m/s
Q.	13	14	15
ANS	v_0	−0.966 m/s	0.66 m/s
Q.	16	17	18
ANS	$4.0\hat{i} + 3.3\hat{j} - 3.3\hat{k}$	(a) 130 m/s (b) 5.02×10^5 J	(a) 0.37 m (b) 6.4 m/s (c) Yes
Q.	19	20	21
ANS	23 m/s	2.0 m/s	−0.27R
Q.	22	23	24
ANS	(a) 5.7 m from the woman (b) 4.2 m, (c) 4.3 m	(a) 5.6 m/s (b) 13 m/s (c) 4.9°	8.2 m/s
Q.	25	26	27
ANS	0.38 m, 1.5 m	−29.6 km/s	65 5 m/s.
Q.	28	29	30
ANS	102 N	(a) $v_{1.4kg} = 14.3$ m/s, $v_{0.28kg} = 17.6$ m/s (b) 347 m	(a) yes (b) decreases by 4800 J
Q.	31	32	33
ANS	31.9 g	$\left(1.70 \dfrac{m^{1/2}}{s}\right)\sqrt{L}$	(b) $x = \dfrac{-M + \sqrt{M^2 + mM}}{m} H$
Q.	34	35	36
ANS	(a) 1.92 m (b) 0.640 m	$m_2 = 2.2$ kg	3.0 m
Q.	37		
ANS	(a) 77.9 m/s (b) 45.0 m/s		

39.11. MULTIPLE-CHOICE ASSIGNMENTS

39.11.1. LEVEL 1

Q. No.	1	2	3	4	5	6	7	8	9
Ans	B	B	A	A	C	A	A	A	D
Q. No.	10	11	12	13	14	15	16	17	18
Ans	A	D	A	D	C	C	B	B	A
Q. No.	19	20	21	22	23	24	25	26	27
Ans	D	A	C	D	A	A	C	A	A
Q. No.	28	29	30	31	32	33	34	35	36
Ans	B	B	A	C	C	C	C	B	C
Q. No.	37	38	39	40	41	42	43	44	45
Ans	A	C	D	D	C	B	B	C	A
Q. No.	46	47	48	49	50	51	52	53	54
Ans	A	C	B	B	C	C	B	C	D
Q. No.	55	56	57	58	59	60	61	62	63
Ans	D	D	C	A	B	D	A	D	C
Q. No.	64								
Ans	D								

39.11.2. LEVEL 2

Q. No.	1	2	3	4	5	6	7	8	9
Ans	A	C	A	A	A	A	A	C	C
Q. No.	10	11	12	13	14	15	16	17	18
Ans	A	C	D	B	D	A	D	D	A
Q. No.	19	20	21	22	23	24	25		
Ans	C	B	C	C	A	B	D		

39.11.3. LEVEL 3

Q. No.	1	2	3	4	5	6	7	8	9
Ans	C	C	C	D	B,C,D	A,C	B	A	D
Q. No.	10	11	12	13	14				
Ans	C	B	D	D	C				

39.11.4. LEVEL 3

39.11.4.1. SECTION A

Q. No.	1	2	3	4	5	6	7	8	9
Ans	C	B	B	A	C	D	A	B	D
Q. No.	10	11	12	13	14	15	16	17	18
Ans	C	B	A	D	B	B	A	A	D
Q. No.	19	20	21	22	23	24	25		
Ans	B	D	B	A	C	B	D		

39.11.4.2. SECTION B

Q.No.	1	2	3	4	5	6	7	8
Ans	A	B	C	D	AD	A	C	A
Q. No.	9	10	11	12	13	14	15	16
Ans	A,C	A	B	A,C	C	4	5	B
Q. No.	17	18	19	20	21	22		

| Ans | B | C | 6.30 | 30.00 | A | 2.09m | | |

39.12. MISCELLANEOUS QUESTIONS

Q. No.	1	2	3	4	5	6	7	8
Ans	C	ABC	C	A	C	D	B	C
Q. No.	9	10	11	12	13	14	15	16
Ans	A	D	A	C	C	D	B	B
Q. No.	17	18	19	20	21	22	23	24
Ans	C	D	C	B	B	F	D	A
Q. No.	25	26	27	28	29	30	31	32
Ans	E	D	B	B	D	C	B	ACD
Q. No.	33	34	35	36				
Ans	AB	D	A	B				

39.13. PRACTICE PROBLEMS

1. The impulse is defined as $J_x = \int_{t_1}^{t_2} F_x dt =$ area under the $F_x(t)$ curve between t_1 and t_2.
   $\Rightarrow J_x = \frac{1}{2}[(6+2) \times 10^{-3}s](1000N) = 4\, N.s$

2. The impulse is defined as $J_x = \int_{t_1}^{t_2} F_x dt =$ area under the $F_x(t)$ curve between t_1 and t_2.
   $\Rightarrow 6.0\, N.s = \frac{1}{2}(4+8) \times 10^{-3}s(F_{max})$
   $\Rightarrow F_{max} = 1.0 \times 10^3 N$

3. Model the object as a particle and the interaction as a collision. The object is initially moving to the right (positive momentum) and ends up moving to the left (negative momentum). By impulse-momentum theorem, we have $p_{fx} = p_{ix} + J_x$
   or $J_x = p_{fx} - p_{ix}$ ... (1)
   As $J_x = F_{avg}\Delta t$
   ∴ From equation (1), $F_{avg} = \frac{p_{fx}-p_{ix}}{\Delta t}$ ... (2)
   Substituting, the values of p_{fx} and p_{ix} and Δt from FIGURE PP3 in above equation, we get
   $F_{avg} = \frac{p_{fx}-p_{ix}}{\Delta t} = \frac{(-2kg.m/s)-(6kg.m/s)}{10\times 10^{-3}s} = -8 \times 10^2 N$
   −ve sigh shows that the direction of force is towards left

6. The impulse is defined as $J_x = \int_{t_1}^{t_2} F_x dt =$ area under the $F_x(t)$ curve between t_1 and t_2.
   $\Rightarrow 6.0\, N.s = \frac{1}{2}(F_{max})(8 \times 10^{-3}s)$
   $\Rightarrow F_{max} = 1.5 \times 10^3 N$

7. Model the glider cart as a particle, and its interaction with the spring as a collision.

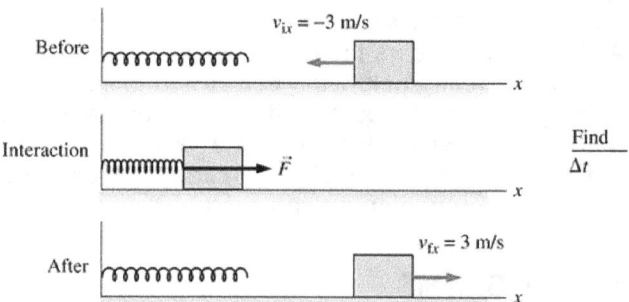

Pictorial representation

Using the impulse-momentum theorem $p_{fx} - p_{ix} = \int F\, dt$, we get
$(0.6\, kg)(3\, m/s) - (0.6\, kg)(3\, m/s)$ = area under force curve $= \frac{1}{2}(36\, N)(\Delta t) \Rightarrow \Delta t = 0.20s$

9. Choose car + rainwater to be the system.
There are no *external* horizontal forces on the car water + system, so the horizontal momentum is conserved

Pictorial Representation

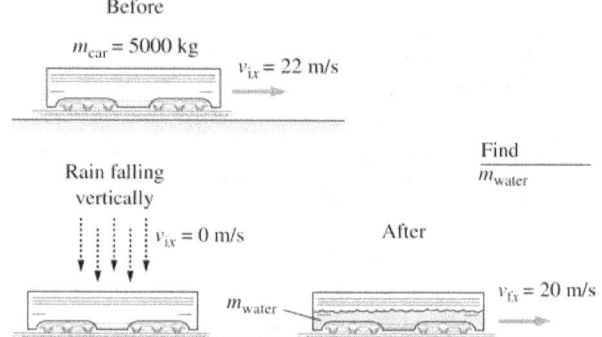

Now, by conservation of momentum, we have
$p_{fx} = p_{ix}.$ .. (1)
Substituting, the values in equation (1), we get
$(m_{car} + m_{water})(20\, m/s)$
$\quad = (m_{car})(22\, m/s) + (m_{water})(0\, m/s)$
$\Rightarrow (5000\, kg + m_{water})(20\, m/s)$
$\quad = (5000\, kg)(22\, m/s)$
$\Rightarrow m_{water} = 5.0 \times 10^2 kg$

12. The throwing of the package is a momentum-conserving action, if the water resistance is ignored. Let A represent the boat and child together, and let B represent the package. Choose the direction that the package is thrown as the positive direction. Apply conservation of momentum, with the initial velocity of both objects being 0.
$p_{initial} = p_{final}$
$\Rightarrow (m_A + m_B)u = m_A v_A + m_B v_B$

$\Rightarrow 0 = m_A v_A + m_B v_B$

$\Rightarrow v_A = -\frac{m_B v_B}{m_A} = -\frac{(5.70\,kg)(10.0\,m/s)}{(24.0\,kg + 35.0\,kg)} = -0.966\,m/s$

The boat and child move in the opposite direction as the thrown package, as indicated by the negative velocity.

14. Since no outside force acts on the two masses, their total momentum is conserved

$m_1 u_1 = m_1 v_1 + m_2 v_2$

$\Rightarrow v_2 = \frac{m_1}{m_2}(u_1 - v_1)$

Now, on substituting given values in above equation you will get the required result.

15. (a) For the initial projectile motion, the horizontal velocity is constant. The velocity at the highest point, immediately before the explosion, is exactly that horizontal velocity, $v_x = v_0 \cos\theta$. The explosion is an internal force, and so the momentum is conserved during the explosion. Let $\vec{v}_s$ represent the velocity of the third fragment.

$\vec{p}_{before} = \vec{p}_{after}$

$\Rightarrow mv_0 \cos\theta\,\hat{\imath}$
$= \frac{1}{3}mv_0 \cos\theta\,\hat{\imath} + \frac{1}{3}mv_0 \cos\theta\,(-\hat{\jmath}) + \frac{1}{3}m\vec{v}_3$

$\Rightarrow \vec{v}_3 = 2v_0 \cos\theta\,\hat{\imath} + v_0 \cos\theta\,\hat{\jmath}$
$= 2(116\,m/s)\cos 60.0°\,\hat{\imath} + (116\,m/s)\cos 60.0°\,\hat{\jmath}$

$\Rightarrow \vec{v}_3 = (116\,m/s)\hat{\imath} + (58.0\,m/s)\hat{\jmath}$

$\Rightarrow v_3 = \sqrt{(116)^2 + (58)^2} \approx 130\,m/s$

Direction of $\vec{v}_3$ above the horizontal is given by

$\tan\alpha = \frac{58}{116} = 0.5 \Rightarrow \alpha = \tan^{-1} 0.5$

$\Rightarrow \alpha \approx 26.6°$

(b) The energy released in the explosion is $K_{after} - K_{before}$. Note that

$v_3^2 = (2v_0 \cos\theta)^2 + (v_0 \cos\theta)^2 = 5v_0^2 \cos^2\theta$

$K_{after} - K_{before}$

$= \left[\frac{1}{2}\left(\frac{1}{3}m\right)(v_0 \cos\theta)^2 + \frac{1}{2}\left(\frac{1}{3}m\right)(v_0 \cos\theta)^2 + \frac{1}{2}\left(\frac{1}{3}m\right)v_3^2\right]$
$- \frac{1}{2}m(v_0 \cos\theta)^2$

$= \left[\frac{1}{2}\left(\frac{1}{3}m\right)(v_0 \cos\theta)^2 + \frac{1}{2}\left(\frac{1}{3}m\right)(v_0 \cos\theta)^2 + \frac{1}{2}\left(\frac{1}{3}m\right)5v_0^2 \cos^2\theta\right]$
$- \frac{1}{2}m(v_0 \cos\theta)^2$

$= \frac{1}{2}mv_0^2 \cos^2\theta\left[\frac{2}{3} + \frac{5}{3} - 1\right] = \frac{2}{3}mv_0^2 \cos^2\theta$

$= \frac{2}{3}(224\,kg)(116\,m/s)^2 \cos^2 60° = 5.02 \times 10^5\,J$

18. (a) At the maximum compression of the spring, the blocks will not be moving relative to each other, and so they both have the same forward speed. All of the interaction between the blocks is internal to the mass-spring system, and so momentum conservation can be used to find that common speed. Mechanical energy is also conserved, and so with that common speed, we can find the energy stored in the spring and then the compression of the spring. Let A represent the 3.0 kg block, let B represent the 4.5 kg block, and let x represent the amount of compression of the spring.

$p_{initial} = p_{final} \Rightarrow m_A u_A = (m_A + m_B)v$

$\Rightarrow v = \frac{m_A}{m_A + m_B} u_A$

$E_{initial} = E_{final}$

$\Rightarrow \frac{1}{2}m_A u_A^2 = \frac{1}{2}(m_A + m_B)v^2 + \frac{1}{2}kx^2$

$\Rightarrow x = \sqrt{\frac{1}{k}[m_A u_A^2 - (m_A + m_B)v^2]}$

$= \sqrt{\frac{1}{k}\frac{m_A m_B}{m_A + m_B} u_A^2}$

$= \sqrt{\frac{1}{850\,N/m}\frac{(3.0\,kg)(4.5\,kg)}{(7.5\,kg)}}(8.0\,m/s)^2 = 0.37\,m$

(b) This is a stationary target elastic collision in one dimension; therefore, we can directly use the formulae for velocities after collision

$v_A = u_A \left(\frac{m_A - m_B}{m_A + m_B}\right) = (8.0\,m/s)\left(\frac{-1.5\,kg}{7.5\,kg}\right) = -1.6\,m/s$

$v_B = u_A \left(\frac{2m_A}{m_A + m_B}\right) = (8.0\,m/s)\left(\frac{6.0\,kg}{7.5\,kg}\right) = 6.4\,m/s$

(c) Yes, the collision is elastic. All forces involved in the collision are conservative forces.

19. Use conservation of momentum in one dimension. Call the direction of the sports car's velocity the positive x direction. Let A represent the sports car, and B represent the SUV. We have $u_B = 0$, and $v_A = v_B$. Solve for u_A.

$p_{initial} = p_{final} \Rightarrow m_A u_A + 0 = (m_A + m_B)v_A \Rightarrow u_A = \frac{m_A + m_B}{m_A} v_A$

The kinetic energy that the cars have immediately after the collision is lost due to negative work done by friction. The work done by friction can also be calculated using the definition of work. We assume the cars are on a level surface, so that the normal force is equal to the weight. The distance the cars slide forward is Δx. Equate the two expressions for the work done by friction, solve for v_A, and use that to find u_A.

$W_{fr} = (K_{final} - K_{initial})_{after\,collision}$

$$= 0 - \tfrac{1}{2}(m_A + m_B)v_A^2$$
$$W_{fr} = F_{fr}\Delta x \cos 180° = -\mu_k(m_A + m_B)g\Delta x$$
$$\Rightarrow -\tfrac{1}{2}(m_A + m_B)v_A^2 = -\mu_k(m_A + m_B)g\Delta x$$
$$\Rightarrow (m_A + m_B)v_A^2 = \mu_k(m_A + m_B)g\Delta x$$
$$\Rightarrow v_A = \sqrt{\mu_k g \Delta x}$$
$$u_A = \tfrac{m_A+m_B}{m_A}v_A = \tfrac{m_A+m_B}{m_A}\sqrt{\mu_k g \Delta x} =$$
$$\tfrac{920\,kg + 2300\,kg}{920\,kg}\sqrt{2(0.80\,m/s^2)(2.8m)} = 23.191\ m/s$$
$$\approx 23\ m/s$$

20. Write momentum conservation in the x and y directions, and kinetic energy conservation. Note that both masses are the same. We allow $\vec{v}_A$ to have both x and y components.

p_x: $mu_B = mv_{Ax}$ $\Rightarrow u_B = v_{Ax}$ ... (1)

p_y: $mu_A = mv_{Ay} + mv_B$
$\Rightarrow u_A = v_{Ay} + v_B$ ... (2)

K: $\tfrac{1}{2}mu_A^2 + \tfrac{1}{2}mu_B^2 = \tfrac{1}{2}mv_A^2 + \tfrac{1}{2}mv_B^2$
$\Rightarrow u_A^2 + u_B^2 = v_A^2 + v_B^2$

Substitute the results from the momentum equations into the kinetic energy equation.

$(v_{Ay} + v_B)^2 + v_{Ax}^2 = v_A^2 + v_B^2$
$\Rightarrow v_{Ay}^2 + v_B^2 + 2v_{Ay}v_B + v_{Ax}^2 = v_A^2 + v_B^2$
$\Rightarrow v_{Ay}^2 + 2v_{Ay}v_B + v_{Ax}^2 = v_A^2 \Rightarrow 2v_{Ay}v_B = 0$
$(\because v_A^2 = v_{Ax}^2 + v_{Ay}^2)$
$\Rightarrow$ either $v_{Ay} = 0$ or $v_B = 0$

Since we are given that $v_B \ne 0$, we must have $v_{Ay} = 0$. This means that the final direction of A is the x direction. Put this result into the momentum equations to find the final speeds.

$v_A = v_{Ax} = v_B = 3.7\ m/s$, $v_B = v_A = 2.0\ m/s$

☞ We can also find the above result by using, Newton's law of restitution instead of conservation of KE.
By Newton's law of restitution, we have
$$v_B - v_A = -e(u_B - u_A)$$
For elastic collision, $e = 1$
$\therefore v_B - v_A = u_A - u_B$
Along X axis: $v_{Bx} - v_{Ax} = u_{Ax} - u_{Bx}$ ... (3)
here, $u_{Ax} = 0$, $u_{Bx} = u_B$, $v_{Bx} = 0$
$\therefore$ from (3), we get $0 - v_{Ax} = 0 - u_B$
or $v_{Ax} = u_B$ (which is same as equation (1))
Along Y axis: $v_{By} - v_{Ay} = u_{Ay} - u_{By}$
here, $u_{By} = 0$, $u_{Ay} = u_A$, $v_{By} = v_B$
$\therefore v_B - v_{Ay} = u_A - 0 \Rightarrow v_B - v_{Ay} = u_A$ ... (4)

From (2) and (4), we have
$v_B - v_{Ay} = v_{Ay} + v_B \Rightarrow v_{Ay} = 0$

22. Since the collision is elastic ($e = 1$), both momentum (in two dimensions) and kinetic energy are conserved. Write the three conservation equations and use them to solve for the desired quantities. You can also solve by using conservation of linear momentum and Newton's law of restitution along x and y direction.

24. Momentum will be conserved in the horizontal direction. Let A represent the railroad car, and B represent the snow.
For the horizontal motion, $u_B = 0$, and $v_B = v_A$. Momentum conservation in the horizontal direction gives the following.
$p_{initial} = p_{final}$ $\Rightarrow$ $m_A u_A = (m_A + m_B)v_A = (m_A + \mu t)v_A$, μ = rate of fall of ice = 3.80 kg/min
$$\Rightarrow v_A = \left(\tfrac{m_A}{m_A + \mu t}\right)u_A$$
$$= \left(\tfrac{4800\,kg}{4800\,kg + \left(\tfrac{3.80\,kg}{min}\right)(60\,min)}\right)8.60\ m/s \approx 8.2\ m/s$$

25. Let A represent the cube of mass M and B represent the cube of mass m. Find the speed of A immediately before the collision, v_A, by using energy conservation.
$$Mgh = \tfrac{1}{2}mu_A^2$$
$$\Rightarrow u_A = \sqrt{2gh} = \sqrt{2(9.8\,m/s^2)(0.35m)} = 2.6\ m/s$$
By Newton's law of restitution, we have
$v_B - v_A = -e(u_B - u_A)$,
For elastic collision, $e = 1$
$\therefore v_B - v_A = u_A - u_B$,
here $u_B = 0$, therefore
$$v_B = u_A + v_A$$
Substitute this relationship into the momentum conservation equation for the collision.
$m_A u_A + m_B u_B = m_A v_A + m_B v_B$ $\Rightarrow$ $m_A u_A = m_A v_A + m_B(u_A + v_A)$
It is also given that $u_B = 0$, $m_A = 2m_B$
$\therefore m_A u_A = m_A v_A + \tfrac{1}{2}m_A(u_A + v_A)$ $\Rightarrow$ $2m_A u_A = 2m_A v_A + m_A(u_A + v_A)$
$\Rightarrow 2u_A = 2v_A + (u_A + v_A) \Rightarrow u_A = 3v_A$
$\Rightarrow v_A = \tfrac{1}{3}u_A = \tfrac{1}{3}\sqrt{2gh}$
$= \tfrac{1}{3}\sqrt{2 \times 9.8\,m/s^2 \times 0.35m} = 0.873\ m/s$
$v_B = u_A + v_A = u_A + \tfrac{1}{3}u_A = \tfrac{4}{3}u_A = 3.492\ m/s$

Each mass is moving horizontally initially after the collision, and so each has a vertical velocity of 0 as they start to fall.

∴ $s = ut + \frac{1}{2}at^2$ gives

$H = \frac{1}{2}gt^2 \Rightarrow t = \sqrt{\frac{2H}{g}}$

Each cube then travels a horizontal distance found by $\Delta x = v_x t$.

$\Delta x_m = v_A \Delta t = \frac{\sqrt{2gh}}{3}\sqrt{\frac{2H}{g}} = \frac{2}{3}\sqrt{hH}$

$= \frac{2}{3}\sqrt{(0.35\ m)(0.95\ m)} = 0.3844\ m \approx 0.38\ m$

$\Delta x_M = v_B \Delta t = \frac{4\sqrt{2gh}}{3}\sqrt{\frac{2H}{g}} = \frac{8}{3}\sqrt{hH}$

$= \frac{8}{3}\sqrt{(0.35\ m)(0.95\ m)} = 1.538\ m \approx 1.5\ m$

26. The interaction between the planet and the spacecraft is elastic, because the force of gravity is conservative. Consider the problem a 1- dimensional collision, with A representing the spacecraft and B representing Saturn. Because the mass of Saturn is so much bigger than the mass of the spacecraft, Saturn's speed is not changed appreciably during the interaction.

Now, by Newton's law of restitution, we have

$v_B - v_A = -e(u_B - u_A)$

For elastic collision, we have $e = 1$, therefore

$v_B - v_A = u_A - u_B$ ... (1)

Here, $u_A = 10.4$ km/s, $u_B = v_B = -9.6$ km/s, $v_A = ?$

Using these values in equation (1), we get

$-9.6\ km/s - v_A = 10.4\ km/s - (-9.6\ km/s)$

or $v_A = -2(9.6\ km/s) - 10.4\ km/s$

or $v_A = -2(9.6\ km/s) - 10.4\ km/s$

$= -29.6$ km/s.

Thus, there is almost a threefold increase in the spacecraft's speed, and it reverses direction.

27. Apply conservation of energy to the motion before and after the collision. Apply conservation of momentum to the collision.

28. First, use conservation of linear momentum to find the speed of the hanging ball just after the collision, after that, apply $\Sigma \vec{F} = m\vec{a}$ to find the tension in the wire. After the collision the hanging ball moves in an arc of a circle with radius $R = 1.35\ m$ and acceleration

$a_{rad} = \frac{v^2}{R}$.

29. The rocket moves in projectile motion before the explosion and its fragments move in projectile motion after the explosion. Apply conservation of energy and conservation of momentum to the explosion.

30. Apply conservation of energy to the motion of the wagon before the collision. After the collision the combined object moves with constant speed on the level ground. In the collision the horizontal component of momentum is conserved.

31. Take the zero of gravitational potential energy to be at the elevation of the pan and let the system include the balance, the beads, and the earth. We can use conservation of energy to find the vertical component of the velocity of the beads as they hit the pan and then calculate the net downward force on the pan from Newton's 2nd law or impulse momentum theorem.

32. Assume that the connecting rod goes halfway through both balls, i.e., the centers of mass of the balls are separated by L. Let the system include the dumbbell, the wall and floor, and the earth. Let the zero of gravitational potential be at the center of mass of the lower ball and use conservation of energy to relate the speeds of the balls to the potential energy of the system. By symmetry, the speeds will be equal when the angle with the vertical is 45°.

33. By symmetry, the center of mass of the empty storage tank should be in the very center, along the axis at a height $y_{t,CM} = H/2$. We can imagine that the entire mass of the tank, $m_t = M$, is located at this point.

The center of mass of the gasoline is also, by symmetry, located along the axis at half the height of the gasoline, $y_{g,CM} = x/2$. The mass, if the tank were filled to a height H, is m; assuming a uniform density for the gasoline, the mass present when the level of gas reaches a height x is $m_g = mx/H$.

(a) The center of mass of the entire system is at the center of the cylinder when the tank is full and when the tank is empty. When the tank is half full the center of mass is below the center. So as the tank changes from full to empty the center of mass drops, reaches some lowest point, and then rises back to the center of the tank.

(b) The center of mass of the entire system is found from

$y_{CM} = \frac{m_g y_{g,CM} + m_t y_{t,CM}}{m_g + m_t}$

$= \frac{\left(\frac{mx}{H}\right)\left(\frac{x}{2}\right) + M\left(\frac{H}{2}\right)}{\left(\frac{mx}{H}\right) + M} = \frac{mx^2 + MH^2}{2mx + 2MH}$

86 CONCEPTS AND PROBLEMS IN PHYSICS

Take the derivative: $\frac{dy_{CM}}{dx} = \frac{m(mx^2+2xMH-MH^2)}{(mx+MH)^2}$

For minimum value of y_{CM}, we have

$\frac{dy_{CM}}{dx} = 0$

$\therefore \quad \frac{m(mx^2+2xMH-MH^2)}{(mx+MH)^2} = 0 \quad \text{or} \quad mx^2 + 2xMH - MH^2 = 0$

or $\quad x = \frac{-M+\sqrt{M^2+mM}}{m} H$

34. (a) Momentum conservation gives,

    $m_R v_R + m_L v_L = 0$

    $\Rightarrow (0.500 \ kg)v_R + (1.00 \ kg)(-1.20 \ m/s) = 0$

    which yields $v_R = 2.40 \ m/s$. Thus, $\Delta x = v_R t = (2.40 \ m/s)(0.800 \ s) = 1.92 \ m$.

    (b) Now we have $m_R v_R + m_L(v_R - 1.20 \ m/s) = 0$, which yields

    $v_R = \frac{(1.2 \ m/s)m_L}{m_L + m_R} = \frac{(1.20 \ m/s)(1.00 \ kg)}{1.00 \ kg + 0.500 \ kg}$
    $= 0.800 \ m/s$

    Consequently, $\Delta x = v_R t = 0.640 \ m$.

35. The velocities of m_1 and m_2 just after the collision with each other are given by equations $v_1 = \frac{2m_2}{m_1+m_2} u_2$,
    $v_2 = \frac{m_2-m_1}{m_1+m_2} u_2$ respectively.

    After bouncing off the wall, the velocity of m_2 becomes $-v_2$. In these terms, the problem requires $v_1 = -v_2$, or

    $v_1 = \frac{2m_2}{m_1+m_2} u_2 = -\frac{m_2-m_1}{m_1+m_2} u_2$

    which simplifies to, $2m_2 = -(m_2 - m_1) \Rightarrow m_2 = \frac{m_1}{3}$.

    With $m_1 = 6.6 \ kg$, we have $m_2 = 2.2 \ kg$.

36. We treat the car (of mass m_1) as a "point-mass" (which is initially 1.5 m from the right end of the boat). The left end of the boat (of mass m_2) is initially at $x = 0$ (where the dock is), and its left end is at $x = 14 \ m$. The boat's center of mass (in the absence of the car) is initially at $x = 7.0 \ m$.

    $\therefore$ the center of mass of the system:

    $x_{CM} = \frac{m_1 x_1 + m_2 x_2}{m_1 + m_2}$
    $= \frac{(1500 \ kg)(14 \ m - 1.5 \ m) + (4000 \ kg)(7 \ m)}{1500 \ kg + 4000 \ kg} = 8.5 \ m$

    In the absence of *external* forces, the center of mass of the system does not change. Later, when the car (about to make the jump) is near the left end of the boat (which has moved from the shore an amount Δx), the value of the CM of system center of mass is still 8.5 m. The car (at this moment) is thought of as a "point-mass" 1.5 m from the left end, so we must have

    $x_{CM} = \frac{m_1 x_1 + m_2 x_2}{m_1 + m_2}$
    $= \frac{(1500 \ kg)(\Delta x+1.5 \ m)+(4000 \ kg)(7 \ m+\Delta x)}{1500 \ kg+4000 \ kg} = 8.5 \ m$

    Solving this for Δx, we find $\Delta x = 3.0 \ m$.

    CM Shift Method

    Suppose shift in the CM of boat is Δx, then

    $\frac{1500(14 - 3 - \Delta x) + 4000(-\Delta x)}{4000 + 1500} = 0$

    or $5500\Delta x = 1500 \times 11$

    or $\Delta x = \frac{1500 \times 11}{5500} = 3 \ m$

38. $mv \cos\theta = (m + m) v'$

    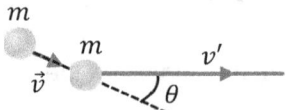

    $\Rightarrow v' = \frac{v}{2} \cos 60° = \frac{v}{4}$

39. $\frac{4x}{(1+x)^2} \times 100$
    $= \frac{4 \times 14}{(1+14)^2} \times 100 = \frac{5600}{225} = 24.9 = 25$

40. Mass of neutron $(m) = 1.67 \times 10^{-27} \ kg$. Speed of neutron $(v) = 1.2 \times 10^7 \ ms^{-1}$. Notice that the mass of deuteron $(M) = 3.34 \times 10^{-27} \ kg = 2m$. If V is the speed of the composite particle, the law of conservation of momentum gives

    $mv = (m + M)V$

    or $\quad V = \frac{mv}{m+M} = \frac{mv}{m+2m}$

    or $\quad V = \frac{v}{3} = \frac{1.2 \times 10^7}{3} = 4 \times 10^6 \ ms^{-1}$.

41. In an elastic collision, both momentum and energy are conserved. Using the two laws, it is easy to see that ($\because M = 2m$), the deuteron will move forward with a speed $2v/3 = \frac{2 \times 1.2 \times 10^7}{3} = 0.8 \times 10^7 \ ms^{-1}$.